Welding and Joining Processes of Metallic Materials

Welding and Joining Processes of Metallic Materials

Guest Editors

Xiaohong Zhan
Jianfeng Wang

MDPI

Basel • Beijing • Wuhan • Barcelona • Belgrade • Novi Sad • Cluj • Manchester

Guest Editors

Xiaohong Zhan
College of Materials Science and Technology
Nanjing University of Aeronautics and Astronautics
Nanjing
China

Jianfeng Wang
College of Materials Science and Technology
Nanjing University of Aeronautics and Astronautics
Nanjing
China

Editorial Office
MDPI AG
Grosspeteranlage 5
4052 Basel, Switzerland

This is a reprint of the Special Issue, published open access by the journal *Materials* (ISSN 1996-1944), freely accessible at: https://www.mdpi.com/journal/materials/special_issues/5D5206HU5F.

For citation purposes, cite each article independently as indicated on the article page online and as indicated below:

Lastname, A.A.; Lastname, B.B. Article Title. *Journal Name* **Year**, *Volume Number*, Page Range.

ISBN 978-3-7258-4321-3 (Hbk)
ISBN 978-3-7258-4322-0 (PDF)
https://doi.org/10.3390/books978-3-7258-4322-0

Contents

About the Editors

Xiaohong Zhan

Prof. Dr. Xiaohong Zhan is currently a Full Professor at the Nanjing University of Aeronautics and Astronautics (NUAA) and the Director of the Laser Welding and Remanufacturing Technology Institute. He received his doctorate in Engineering in Materials Processing Engineering from the Harbin Institute of Technology in 2008. From 2009 to 2012, he has held a concurrent position at the Commercial Aircraft Corporation of China Ltd. (COMAC) as the Deputy Director of the "Advanced Manufacturing Technology R&D Center for Civil Aircraft". During 2016–2017, he conducted research as a state-sponsored visiting scholar at The Ohio State University (OSU), USA, under the guidance of Prof. Alan Luo (member of the National Academy of Engineering), focusing on the metallurgical theory of light alloys. He has published over 200 papers of archival significance in influential SCI journals such as *Composites Part B: Engineering, Additive Manufacturing, Journal of Materials Science & Technology*, etc., and he has over 80 authorized invention patents. He has led over ten national-level projects, including four grants from the National Natural Science Foundation of China (NSFC), and over 30 equipment development projects in aerospace and related fields. His accomplishments have been honored with prestigious awards such as the First Prize of the Science and Technology Award of Jiangsu Province, the Second Prize of the National Defense Technology Invention Award, and multiple other provincial-/ministerial-level accolades. He currently holds leadership roles as the Deputy Director of the Youth Committee of the Welding Society (Chinese Mechanical Engineering Society), an Executive Council Member of the Chinese Welding Association, and an Executive Council Member of the Jiangsu Welding Society.

Jianfeng Wang

Jianfeng Wang is currently an Associate Professor at the College of Materials Science and Technology, Nanjing University of Aeronautics and Astronautics, China. He earned his bachelor's degree (2013), master's degree (2015), and PhD degree (2019) from the State Key Laboratory of Advanced Welding and Joining, Harbin Institute of Technology (HIT) in China. He has published 68 papers of archival significance, including 32 as a first author and 16 as a corresponding author. In total, 56 of his journal papers appear in influential SCI journals, such as *Weld. J.*, *Virtual Phys. Prototy.*, *J. Mater. Sci. Technol.*, *Mater. Sci. Eng. A*, *J. Mater. Process. Tech.*, etc. His paper in *J. Manuf. Process.* (2019, 37: 563-577) has been recognized as a "Key Scientific Article" by Advances in ENGINEERING (Canada), the World's leading source of engineering research news. He has applied/authorized 36 national invention patents. He has also been granted the European Union's Horizon 2020 Research and Innovation Programme under the Marie Sklodowska-Curie Actions (2022). He is devoted to fundamental research activities involving laser welding, additive manufacturing with an emphasis placed on microstructure control via ultrasonic waves, addressing issues covering arc behavior, metal transfer, and the microstructure modification of weld metal. He is a Committee Member of the Laser Processing Committee of the Chinese Optical Society and serves as the Associate Editor for the Board of the Advance in *Materials Science*.

Preface

The last two decades have seen an intensive improvement in material welding, joining, and additive manufacturing methods, enabling the weight reduction and high functionalization of multi-material structures. Today, it is possible to fabricate large-sized and thin-walled structures made of different types of metallic alloys with a more complicated geometry of reinforcement, including nanoparticles or precipitated phases. The advanced welding, joining, and additive manufacturing processes of complex structures allows for the development of new technologies, with recent advances in manufacturing techniques further maximizing functionality while retaining the original characteristics of the structure.

The main purpose of this Special Issue is to collect research on the advanced processes in material welding, joining, and additive manufacturing aspects. As such, a fundamental understanding of the process–microstructure–property relationships is necessary. The Special Issue was a success, with a total of eight original research studies published after peer review. Different topics were discussed within this Special Issue, including the hot-cracking mechanism of laser welding of aluminum alloys; residual stress in welding; the microstructure and mechanical properties of narrow gap welding and electron beam welding; and wire arc additive manufacturing. The last two decades have also witnessed an accelerated development of welding, joining, and additive manufacturing processes that enable the high-strength and defect-free welding of joints for materials with various physical properties. This Special Issue represents an excellent opportunity for researchers around the world to share different aspects of their work and report results related to this topic.

We thank to all the authors and peer reviewers for their valuable contributions to the Special Issue "Welding and Joining Processes of Metallic Materials".

Xiaohong Zhan and Jianfeng Wang
Guest Editors

materials

MDPI

Article

Investigation of Arc Stability in Wire Arc Additive Manufacturing of 2319 Aluminum Alloy

Qiyu Gao *, Feiyue Lyu, Leilei Wang and Xiaohong Zhan *

College of Materials Science and Technology, Nanjing University of Aeronautics and Astronautics, Nanjing 211106, China; feiyue_lv@163.com (F.L.); wangll@nuaa.edu.cn (L.W.)
* Correspondence: qiyugao_nuaa@126.com (Q.G.); xiaohongzhan_nuaa@126.com (X.Z.)

Abstract: Wire Arc Additive Manufacturing (WAAM) technology, known for its low equipment and material costs, high material utilization, and high production efficiency, has found extensive applications in the fabrication of key components for the aerospace and aviation industries. The stability of the arc is crucial for the WAAM process as it directly affects the forming of the parts. In this study, the monitoring data of electrical signals and arc morphology during the WAAM process of 2319 aluminum alloy were investigated using a high-speed camera system and current/voltage acquisition system. By analyzing the current and voltage signals, as well as the arc imaging results, the influence of arc stability on the forming of the cladding layer was studied. The experimental results indicated that when both current and voltage exhibit regular periodic fluctuations, this manifests as a stable short-circuit droplet transition form, while sudden changes in these signals represent abnormal droplet transition forms. The adaptability of the process directly influenced the arc shape, thereby affecting the forming of the cladding layer. Under the process parameters of welding speed of 240 cm/min and wire feeding speed of 6.5 m/min, it was observed that the current signal exhibited a tight state and the variance of the arc width was minimized. This indicated that at a higher wire feeding speed, the droplet transfer frequency was increased. Under these process parameters, the arc output was more stable, resulting in a uniform metal coating.

Keywords: 2319 aluminum alloy; wire arc additive manufacturing; arc stability; current and voltage signals; arc morphology

1. Introduction

The material 2319 aluminum alloy is highly regarded in the aerospace industry due to its excellent thermal properties and high strength-to-weight ratio, making it ideal for the production of variable cross-section thin-walled components [1–3]. Traditional manufacturing methods of variable cross-section aluminum alloy thin-walled structures involving mechanical machining are associated with drawbacks such as material wastage and extended production timelines [4,5]. In contrast, Wire Arc Additive Manufacturing (WAAM) offers a compelling solution with its superior material utilization, adaptability for intricate component geometries, and enhanced manufacturing flexibility [6,7]. WAAM is an additive manufacturing process developed based on welding technology, where metal wire is heated and melted by an electric arc and then deposited layer by layer to construct a three-dimensional structure. Compared to laser and electron beam additive manufacturing technologies, WAAM offers high deposition rates, low equipment and material costs, large-scale manufacturing capabilities, and minimal post-processing requirements. Consequently, it demonstrates significant application advantages in the aerospace, shipbuilding, automotive, and energy industries. Among them, a controlled short arc process with wire retraction, namely, cold metal transition (CMT) by the Fronius company, has the advantages of a stable arc shape, reduced spatter, and low heat input. The WAAM system based on CMT technology is used to manufacture Al-Mg alloy walls, exhibiting good tensile and impact fracture properties. The results show that the Al-Mg alloy produced by WAAM-CMT has superior properties compared to traditional casting alloys [8]. Li et al. used WAAM-CMT to prepare Al-5.49Cu-0.4Mn-0.29Cd alloy in order to obtain a uniform microstructure and improve its mechanical properties. Based on the experimental results and theoretical analysis, the strengthening mechanism and fracture mechanism of the CMT-WAAM aluminum alloy were revealed [9]. It can be seen that this technology can better control shape quality and improve manufacturing efficiency [10].

Arc stability refers to the degree to which the arc maintains a stable formation during the welding process without experiencing phenomena such as arc interruption. A stable arc has the advantages of smooth droplet transition, small arc length variation, less short-circuit splashing, beautiful weld formation, and high welding quality. On the contrary, if the arc formation is unstable, causing severe splashing during droplet transition, poor weld formation, and easy formation of defects such as porosity or slag inclusion, the welding quality will be low [11,12]. The relationship between process parameters and arc morphology has become a major area of interest for researchers aiming to optimize WAAM process schemes. By establishing a comprehensive understanding of this correlation, it becomes possible to improve the overall performance and efficiency of WAAM processes.

By continuously monitoring key manufacturing parameters in WAAM, such as arc stability, current, voltage, temperature, and others, potential manufacturing defects can be promptly detected and corrected [13]. This ensures that the final products meet the expected levels of quality and performance. By monitoring the stability of the arc and making adjustments as necessary, process consistency is ensured, thereby reducing variations in the finished products [14]. This is crucial for meeting stringent quality standards and specifications. Therefore, process monitoring proves to be advantageous in gaining control over the forming quality in the additive manufacturing process, mitigating potential risks and reducing time, research, and development expenses [15,16].

Hauser et al. [17] investigated the WAAM of AW4043/AlSi5(wt.%) under various welding torch lead angles using high-speed imaging to observe welding phenomena, revealing that specific lead angles and interlayer temperatures caused fluctuation effects in the deposited structures, which could be detected at the arc frequency of the process. They concluded that a pushing WAAM process with a slightly tilted lead angle can mitigate these effects. Zhu et al. [18] monitored the arc current and voltage signals during the WAAM process to identify droplet transfer patterns. By calculating the probability density distribution, power spectrum distribution, and energy gradient of time-domain synchronous signals, the characteristics of metal droplet transfer and energy transfer were studied. Ke et al. [19] utilized high-speed imaging to capture the droplet transfer process, which was then employed to establish a periodic droplet transfer model. Doumanidis et al. [20] investigated the influence of welding speed and wire feed rate on the geometric dimensions of molten droplets in the WAAM process. By varying the process parameters, they were able to control the size of the molten droplets and achieve precise control over the macroscopic morphology of the WAAM components.

In summary, many researchers and experts have found that process parameters have a significant impact on arc stability [21], manifested in changes in the geometry of the deposition layer [14], arc morphology [22], and voltage/current signals [23]. However, the basic physical factors that affect arc stability and their mechanisms of action need to be further studied. Therefore, this study utilized a high-speed camera system and current/voltage acquisition system to collect monitoring data on electrical signals and arc morphology and systematically analyzed the current and voltage signals as well as arc imaging results. The aim was to investigate the effects of welding speed and wire feeding speed on surface forming quality, electrical signal changes, droplet transition behavior, and arc morphology during the WAAM process; screen the influencing factors of arc stability; optimize parameters; and achieve stable and high-quality cladding layer formation.

2. Materials and Methods

The WAAM equipment and signal acquisition system are shown in Figure 1, including a Trans Plus Synergic 4000 CMT welding machine, KUKA KR10R 1420 C4 robot, VR1500 4R/F++ROBOTER wire feeding mechanism, current/voltage signal acquisition system, and AcutEye high-speed camera system (developed by Hunan Kotianjian Optoelectronics Technology Co., Ltd. (Changsha, China), equipped with a German Optronis high-speed camera, matching laptop industrial computer, and AcutEye software v4.0). The high-speed camera was positioned approximately 0.5 m away from the workpiece, angled at 45° relative to the horizontal worktable. Additionally, a filter lens was mounted on the front of the high-speed camera, which leveraged the light source emitted by the laser emitter to capture the arc morphology during the printing process. The electrical signal monitoring system used consisted of a current/voltage signal acquisition system, a signal processing system, and a computer. The current/voltage signal acquisition system was connected to the CMT welding machine to dynamically collect the current and voltage values output by the welding machine in real time. The acquisition system included a sampling amplification circuit and a data acquisition card. This data acquisition card had four differential input analog channels, with the core being a 12-bit A/D converter ADS7800, and a single acquisition time of 5000 ms. The analog channel of the acquisition card was sequentially connected to the arc voltage and welding current signals. The collected signals were transmitted through cables to the signal processing system for waveform analysis and, finally, the waveform display software in the computer was used to monitor the real-time changes in current and voltage during the arc additive process. This signal processing system could perform synchronous sampling and waveform analysis on multi-channel signals and calculate and process time-domain parameters such as the average value, effective value, and peak value of the signal. In addition, the signal processing system could achieve functions such as wavelet denoising, wavelet analysis, power spectrum estimation, and optimal feature subset extraction of signal sources. The electrical signals collected during the arc feeding process were current and arc voltage signals, where the voltage signal was usually proportional to the arc length. The formation of the arc and the transition of the droplet caused periodic changes in the arc length and resistance in the feeding circuit, thereby affecting the periodic changes in current/voltage in the feeding circuit. Therefore, when there were phenomena such as arc breaking and extinguishing during the additive process, the fluctuation of arc voltage and current could indirectly reflect the stability of the arc. The chemical compositions of AA2319 wire and AA2219 substrate are presented in Table 1, which were obtained based on Inductively Coupled Plasma Emission Spectrometer (ICP) experimental results.

Table 1. Chemical compositions of AA 2319 wire and AA 2219 substrate in wt.%.

Element	Si	Fe	Cu	Mn	Mg	Zn	Ti	Zr	V	Al
2319 wire	0.0420	0.0672	6.7381	0.2283	0.0066	0.0034	0.1132	0.1182	0.0845	Bal.
2219 substrate	0.0671	0.0823	4.1335	0.4663	0.0166	0.0559	0.0349	0.0003	0.0098	Bal.

The process parameters (welding machine settings) for WAAM are shown in Table 2. This method adopted a synergic mode to adjust the wire feeding speed, arc voltage, and current. That is, by selecting the wire feeding speed, the welding machine automatically set the arc voltage and current accordingly. The welding gun was perpendicular to the substrate, the dry elongation of the welding wire was set to 10 mm, and the welding path of the welding gun was S-shaped. In the CMT WAAM process, the wire underwent periodic retraction, facilitating the detachment of the droplet from the wire. This ensured that the droplet transfer occurred with an extremely low current value, enabling the continuous alternation between "cold" and "hot" states during the additive process. In addition, to prevent impurities from entering the molten pool during the WAAM process, resulting in material oxidation or increased porosity defects, argon gas with a purity of 99.99% [24,25]

was used for local protection, and the shielding gas nozzle diameter was 30 mm. This gas was output coaxial through the welding gun, with a gas flow rate of 23 L/min. The target size of the WAAM test block was 130 mm × 20 mm × 10 mm (length × width × height). The test block was printed in 8 layers and a single weld bead with one layer, with an average layer height of 1.5 mm.

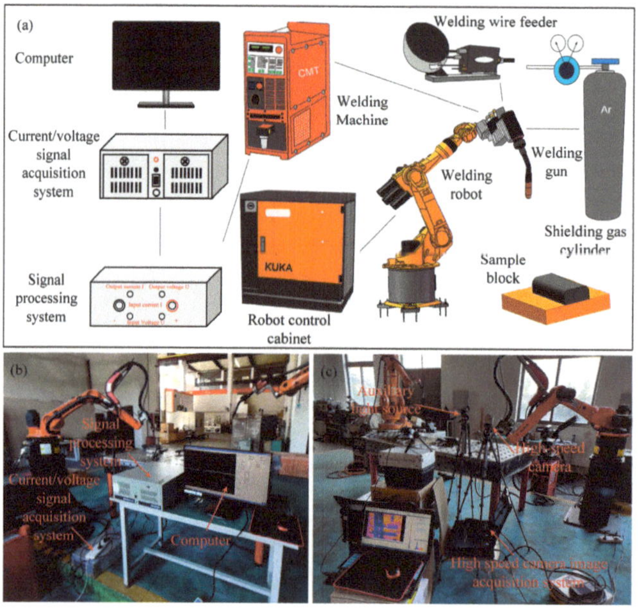

Figure 1. Types of equipment of WAAM experiments: (**a**) model diagram; (**b**) signal acquisition system; (**c**) high-speed camera image acquisition system.

Table 2. Summary of AA2319 WAAM process parameters.

Experimental Case	Current (I) (A)	Voltage (U) (V)	Wire Feeding Speed (v_f) (m/min)	Welding Speed (v_w) (cm/min)
1	90	10.7	5.5	120
2	90	10.7	5.5	180
3	90	10.7	5.5	240
4	120	13.4	6.0	120
5	120	13.4	6.0	180
6	120	13.4	6.0	240
7	131	14.6	6.5	120
8	131	14.6	6.5	180
9	131	14.6	6.5	240
10	140	15.0	7.0	120
11	140	15.0	7.0	180
12	140	15.0	7.0	240

3. Results and Discussion

3.1. Macroscopic Morphology

Figure 2 shows the surface forming quality of the WAAM samples with a welding speed of 120~240 cm/min and wire feeding rates of 5.5 m/min, 6.0 m/min, 6.5 m/min, and 7.0 m/min, respectively. As shown in Figure 2a, when the wire feeding rate was 5.5 m/min, the formation of the upper surface of the cladding layer was poor. When the welding speed was 120 cm/min, the ripples on the surface of the cladding layer were irregular and the ripples in the middle of the weld bead were pronounced but lacked continuity. When the welding speed was 180 cm/min, the surface of the cladding layer was non-uniform, with intermittent ripples present in the middle area of the cladding layer surface. When the welding speed was 240 cm/min, the ripples on the surface of the cladding layer were highly irregular, and serious intermittent ripples were present in the middle area of the cladding layer surface.

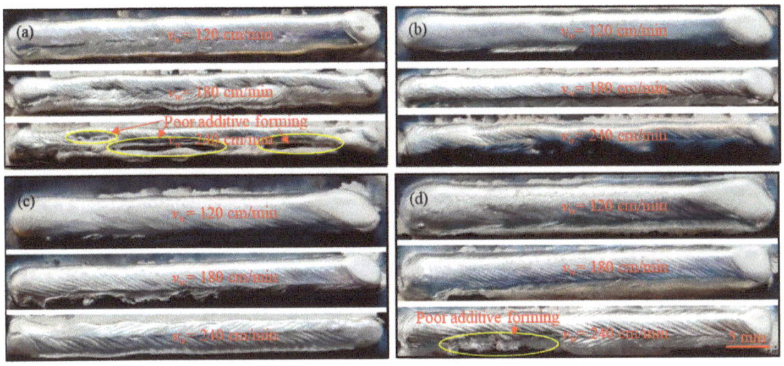

Figure 2. The influence of welding speed changes on AA2319 WAAM forming quality: (**a**) wire feeding speed 5.5 m/min; (**b**) wire feeding speed 6.0 m/min; (**c**) wire feeding speed 6.5 m/min, (**d**) wire feeding speed 7.0 m/min.

As shown in Figure 2b, at a welding speed of 120 cm/min, the continuity of the ripples on the upper surface of the cladding layer was enhanced, with no evident discontinuities or irregularities midway. However, the width of the top surface of the cladding layer increased compared to the condition with a wire feed speed of 5.5 m/min. At a welding speed of 180 cm/min, the surface of the cladding layer exhibited smooth and aesthetically pleasing welding ripples, indicating optimal heat input under these process parameters. This resulted in uniform metal deposition within the cladding layer. At a welding speed of 240 cm/min, the surface of the cladding layer became non-uniform, with fewer porosities and solidified droplets observed in the central region of the cladding layer.

As illustrated in Figure 2c, with a wire feed speed of 6.5 m/min, the surface of the cladding layer appeared smooth and glossy, and the consistency of the weld bead ripple was well maintained. When the welding speed was set to 120 cm/min, the width of the additive cladding layer increased further. However, at a welding speed of 180 cm/min, a slight collapse was observed at the right end of the cladding layer in the arc termination zone, likely due to a low arc stopping current or a short arc stopping time. A welding speed of 240 cm/min resulted in relatively high side roughness on the cladding layer.

As depicted in Figure 2d, with a wire feed speed of 7.0 m/min, the upper surface of the cladding layer remained smooth and flat, and the cladding layer width increased further. This was attributed to the higher wire feed speed resulting in an increased matching current and voltage of the welding machine, which, in turn, increased the heat input during the arc deposition process. This elevated heat input promoted extensive spreading of the molten metal and led to a wider cladding layer upon rapid solidification. At a welding speed of 120 cm/min, a noticeable collapse was observed at the start/stop positions of the arc. When the welding speed was 180 cm/min, a slight collapse occurred at the arc termination point on the right end, which could be attributed to a low arc stopping current or a short arc stopping time. At a welding speed of 240 cm/min, the cladding layer exhibited relatively high side surface roughness.

3.2. Changes in Electrical Signals

The changes in voltage/current signals with time at different layers at a welding speed of 120 cm/min and a wire feeding rate of 5.5 m/min are shown in Figure 3. The results of the 1000~15,000 ms WAAM process indicated that there were no abnormalities in the current and voltage signals. During the WAAM process, the wire was sent out to make contact with the substrate and start arc striking. At this point, the arc melted the substrate and formed a molten pool. Next, the wire melted and formed droplets, which quickly transitioned into the molten pool. At this point, the corresponding electrical signals during the addition process underwent periodic changes.

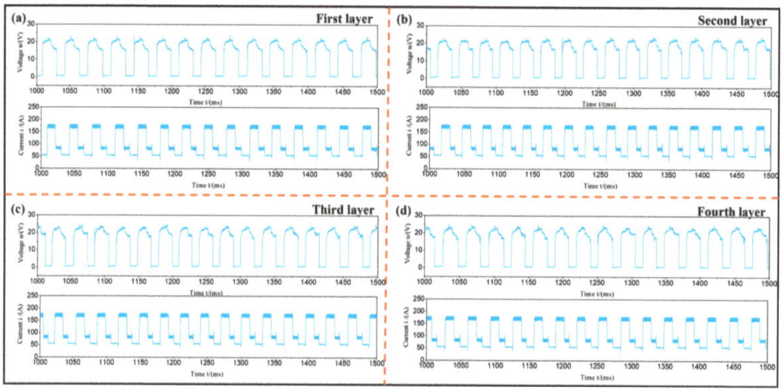

Figure 3. The variation of voltage/current signal with time at different layers at a welding speed of 120 cm/min and a wire feeding rate of 5.5 m/min: (**a**) first layer; (**b**) second layer; (**c**) third layer; (**d**) fourth layer.

Figure 4 shows the relationship between the arc droplet transfer process and the current/voltage waveform during the WAAM process. As shown in Figure 4a, the droplet underwent a short-circuit transition and the current value rapidly dropped below 60 A, while the voltage value dropped sharply to 0.5–0.8 V. Subsequently, the wire feeding speed reversed, the wire was pulled back, and the current and voltage were almost zero. As shown in Figure 4b, the current/voltage value immediately increased and the arc was reignited. As shown in Figure 4c,d, within 40 ms, the current/voltage values remained basically unchanged, that is, this was the peak time of the current. The end of the wire continued to heat up, causing the droplet to grow. Meanwhile, under the action of the electromagnetic contraction force, the root of the droplet shrank. As shown in Figure 4e, during the arc closing phase, the current and voltage values slightly decreased. The welding wire was gradually fed downward and the arc width gradually decreased. When the welding wire came into contact with the molten pool, the droplet was once again sent into the molten pool through a short-circuit transition.

By analyzing the current/voltage signal waveform of the intercepted 1000~15,000 ms WAAM process, it was found that there were abnormal phenomena in the current/pressure signal of the third layer. The current and voltage waveforms were continuously monitored up until 1800 ms. Through observation, it was found that the voltage of the third layer was 0 V and the current reached 300 A. This was because the wire was in contact with the substrate at this time, and the arc voltage was 0 V. However, the current continued to reach the substrate through the wire, generating an instantaneous current of up to 300 A, as shown in Figure 5. The reason for this phenomenon was that during the arc feeding process, the wire continuously extended and retracted, leading to periodic fluctuations in current and voltage signals. The wire did not always interact with the workpiece. When the two were tightly connected and could not be separated, this could lead to wire-sticking behavior. This led to a tight connection between the cathode and anode without generating a potential difference. Therefore, if the arc voltage value was 0 V and the resistance value of the wire was low, the internal current value immediately increased to 300 A. Consequently, in this process combination, the welding speed was higher and the wire feeding speed was lower, resulting in extremely low heat input values in the addition process. The wire could not be transported by jet but could only enter the molten pool through short-circuit transportation. At the same time, due to low heat input, the wire could not fully melt, resulting in poor spreading of the cladding layer and smaller overlapping areas between the weld beads. This led to more voids in the middle of the cladding layer.

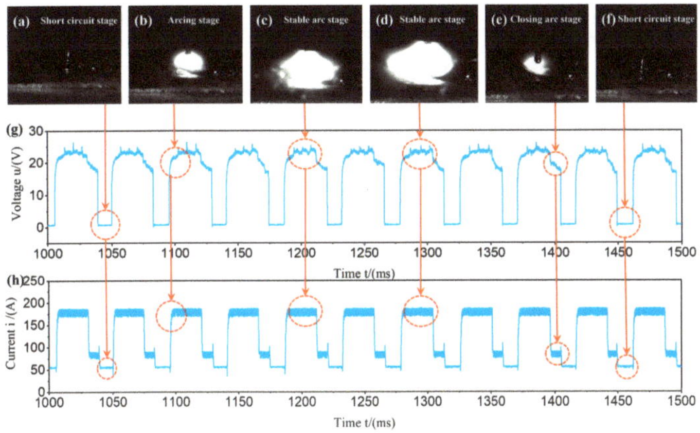

Figure 4. The relationship between the arc droplet transfer process and the current/voltage waveform in WAAM: (**a**) short circuit stage; (**b**) Arcing stage; (**c**) stable arc stage; (**d**) stable arc stage; (**e**) closing arc stage; (**f**) short circuit stage; (**g**) variation of voltage signal with time; (**h**) variation of current signal with time.

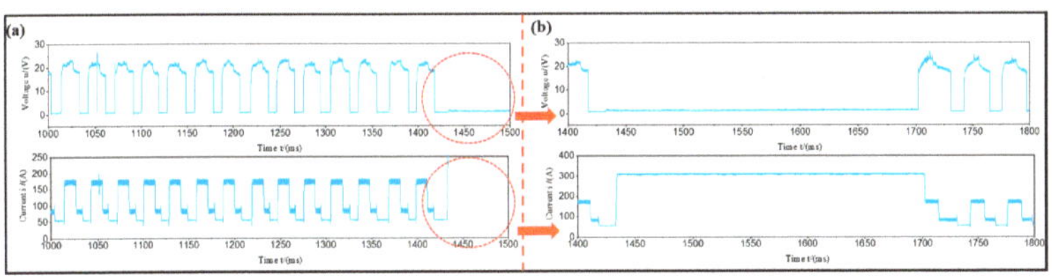

Figure 5. The variation of voltage/current signal with time at the third layer at a welding speed of 240 cm/min and a wire feeding rate of 5.5 m/min: (**a**) at 1000 ms~1500 ms; (**b**) at 1400 ms~1800 ms.

3.3. Droplet Transfer

Real-time monitoring of the droplet morphology and droplet drop process during the arc addition process of 2319 aluminum alloy was carried out through high-speed cameras. In the closing arc stage, when the maximum droplet was formed at the tip of the welding wire, the maximum diameter size of the droplet was measured three times in each experiment, and the average value was taken to analyze the influence of welding speed and wire feeding speed on the shape and size of the droplet, as shown in Figure 6.

During the WAAM process, the droplet transfer remained in the form of short-circuit transfer under different process parameters, and the droplet contacted the molten pool before it grew, forming a short circuit. The shape of the droplet remained basically circular or nearly circular, and the overall width of the droplet was distributed from 1.3 mm to 2.2 mm. Due to the impact of arc recoil force and the surface tension of the wire, some of the droplet sizes decreased in the vertical direction and increased in the horizontal direction, forming an elliptical droplet. Therefore, the overall width and size distribution of the droplet were relatively wide, and the droplet size changed significantly under a single parameter. In addition, from Figure 6, we can see that the droplet size under all parameters was greater than the wire size. This was because there was surface tension at the contact surface between the wire and the droplet, which was related to the material, wire diameter, gas medium, and temperature. The direction of surface tension was opposite to gravity and had a certain limiting effect on the drop of molten droplets. Only when the droplet continued to grow until the size of the droplet reached a certain level, allowing the gravity of the droplet to overcome resistance, could it enter the molten pool.

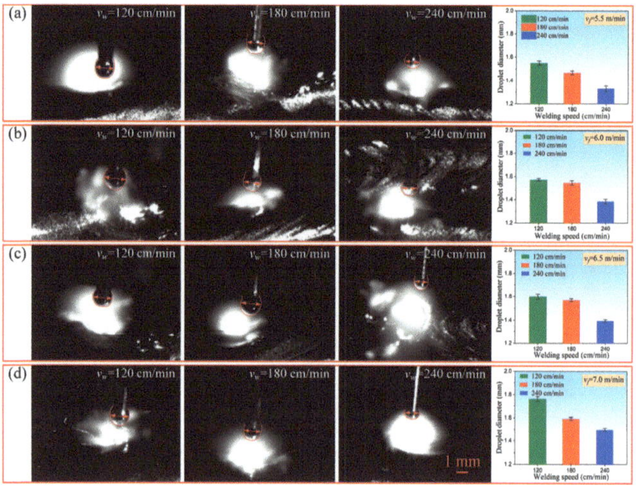

Figure 6. Droplet transfer behavior during arc additive process: (**a**) wire feeding speed 5.5 m/min; (**b**) wire feeding speed 6.0 m/min; (**c**) wire feeding speed 6.5 m/min; (**d**) wire feeding speed 7.0 m/min.

As depicted in Figure 6, there was a negative correlation between the droplet size and welding speed. When the wire feeding speed remained constant, increasing the welding speed resulted in smaller droplet sizes. This was because a faster welding speed led to faster movement of the welding gun, resulting in reduced heat input to the molten pool per unit time and area. Consequently, there was a decrease in the amount of metal filling entering the molten pool. As the number of printing layers increased, the distance between the welding gun and the upper surface cladding layer also increased, causing the arc length to increase. This led to noticeable heat dissipation in the arc column area. Simultaneously, as the arc heat and resistance heat decreased, the heat that promoted droplet growth also decreased. Consequently, the melting amount of the wire gradually reduced, resulting in smaller droplet sizes during the formation process.

3.4. Evolution of Arc Morphology

Observation and measurements were conducted on the region with the maximum arc expansion within a certain period of time, as shown in Figure 7. During the WAAM process, the arc morphology was close to a bell shape. When the wire feeding speed was low, the welding gun lifted by 1.5 mm per layer, while the amount of cladding per layer was less than 1.5 mm, resulting in an increasing distance between the welding gun and the upper surface of the cladding layer. The arc first diverged and then converged due to the electromagnetic contraction force, ultimately forming a bell-shaped arc. When the distance between the welding gun and the substrate was very long, the arc was elongated and the arc voltage increased. In order to maintain a stable voltage value, the welding machine slightly increased the current value, resulting in an increase in the amount of metal melted per unit time and an increase in the amount of cladding per layer. Therefore, the distance between the welding gun and the substrate returned to the original value, and the arc was sprayed radially, with a larger arc width.

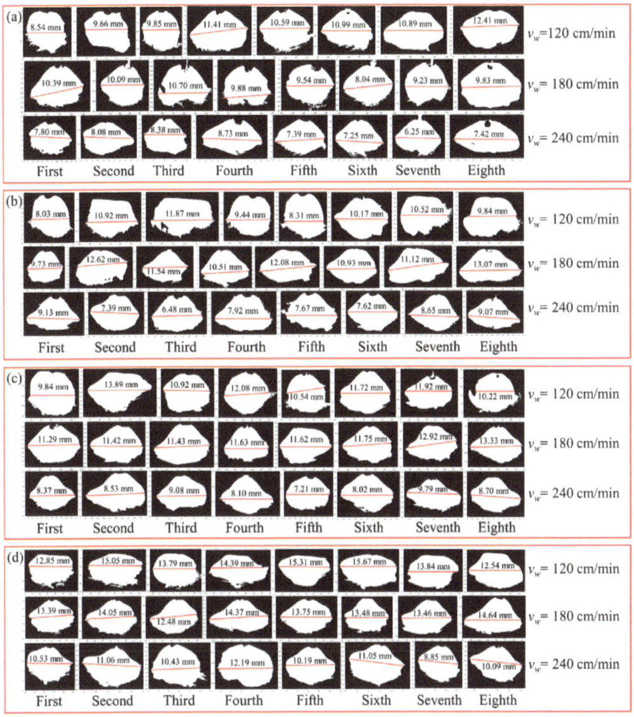

Figure 7. Effect of welding speed variation on arc morphology at different layer heights: (**a**) wire feeding speed 5.5 m/min; (**b**) wire feeding speed 6.0 m/min; (**c**) wire feeding speed 6.5 m/min; (**d**) wire feeding speed 7.0 m/min.

The maximum width of the extracted arc contour was measured using Image Pro Plus software v6.0, with three measurements taken each time. The average value of the results is shown in Figure 8, which illustrates the impact of different layer height/welding speed variations on the arc. When the given wire feeding speed was consistent, the arc width showed a decreasing trend as the welding speed increased from 180 cm/min to 240 cm/min. This was because as the welding speed increased and the matching current and voltage values of the welding machine were constant, the heat input per unit time in a fixed size area decreased and the mass of droplets entering the molten pool per unit time decreased. When the single-layer cladding amount was less than the lifting amount of each layer of the welding gun, the distance between the welding gun and the upper surface of the cladding layer gradually increased, the arc length increased, and the width decreased. The end of the arc converged due to the electromagnetic contraction force, ultimately forming a bell-shaped arc.

Figure 9 shows the influence of wire feeding speed changes on the average and variance of arc width. The method of taking this average value was to measure the average arc width of the eight layers of additive under each process parameter. The variance was calculated to evaluate the difference between the arc width of each layer and the average arc width of the eight layers under this parameter, and the stability of the arc under this parameter was judged through data volatility. If the variance value of the arc width was small, this meant that the data volatility was small, and the arc had good stability during the multi-layer printing process. However, the variance value of the arc width was relatively large, which meant that the data volatility was high, indicating that the stability of the arc was poor during the multi-layer printing process.

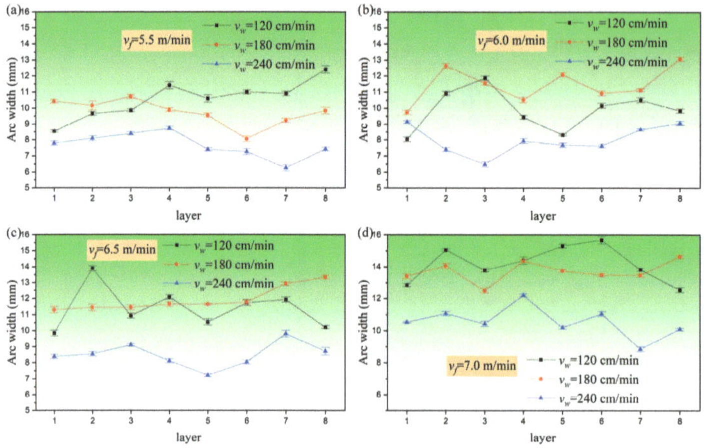

Figure 8. Curve plot of the effect of welding speed changes on arc morphology at different layer heights: (**a**) wire feeding speed 5.5 m/min; (**b**) wire feeding speed 6.0 m/min; (**c**) wire feeding speed 6.5 m/min; (**d**) wire feeding speed 7.0 m/min.

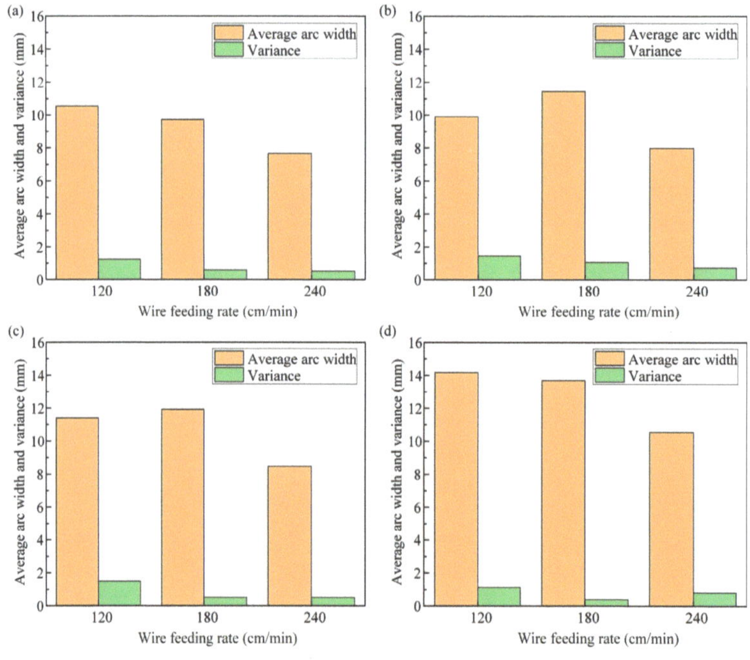

Figure 9. Histogram of the influence of welding speed changes on the average and variance of arc width: (**a**) wire feeding speed 5.5 m/min; (**b**) wire feeding speed 6.0 m/min; (**c**) wire feeding speed 6.5 m/min; (**d**) wire feeding speed 7.0 m/min.

In summary, to obtain thin-walled structures, when the welding speed was maintained at 240 cm/min and the wire feeding speed was 6.5 m/min, the variance value of the arc width was the lowest. This meant that during the process of printing eight layers of arc width under this parameter, the fusion amount of each layer matched the lifting amount of each layer of the welding gun, and the arc length and width basically did not change. Therefore, the stability of the printed arc was the highest under this process parameter.

4. Conclusions

(1) The process parameters had a critical impact on the quality of the coating layer formation. At a wire feed speed of 6.5 m/min, the resulting coating exhibited a smooth and glossy surface finish, with excellent weld continuity and wave uniformity.

(2) In the current process window, the short circuit phenomenon of the arc triggered periodic changes in current and voltage signals, that is, the arc formed, and the wire was fed forward until a droplet short circuit was formed. At this moment, the wire feeding speed was reversed, the wire was pulled back, and the current and voltage were almost zero. After the next open circuit was formed, the arc reignited, the welding wire was stopped and sent forward again, and the droplet transition began again. When the voltage suddenly dropped to 0 V and the current reached 300 A, this indicated that the welding wire was embedded in the molten pool or there was insufficient heat input, resulting in a lack of fusion and an increase in voids within the cladding layer.

(3) There was a negative correlation between droplet size and welding speed, where higher welding speeds resulted in reduced heat input, leading to smaller droplets. In addition, with the increase in the number of cladding layers, the distance between the welding gun and the substrate further reduced heat transfer, resulting in a decrease in droplet size.

(4) When the welding speed was maintained at 240 cm/min and the wire feeding speed was 6.5 m/min, the variance of the arc width was the smallest. Under these process parameters, the stability of the welding arc was the highest.

Author Contributions: Conceptualization, Q.G. and X.Z.; methodology, F.L.; validation, F.L., Q.G. and L.W.; formal analysis, Q.G.; investigation, Q.G.; resources, X.Z.; data curation, F.L.; writing—original draft preparation, Q.G.; writing—review and editing, F.L.; supervision, L.W.; funding acquisition, X.Z. and Q.G. All authors have read and agreed to the published version of the manuscript.

Funding: This research was funded by the Defense Industrial Technology Development Program, grant number JCKY2020605C006, and the Jiangsu Funding Program for Excellent Postdoctoral Talent, grant number 2023ZB687.

Institutional Review Board Statement: Not applicable.

Informed Consent Statement: Not applicable.

Data Availability Statement: The original contributions presented in the study are included in the article, further inquiries can be directed to the corresponding authors.

Conflicts of Interest: The authors declare no conflicts of interest.

References

1. Fu, R.; Lu, W.; Guo, Y.; Lei, H.; Cui, Y.; Wang, J.; Gao, D.; Wang, J.; Liu, C. Achieving high strength-ductility of Al-Zn-Mg-Cu alloys via hot-wire arc additive manufacturing enabled by strengthening precipitates. *Addit. Manuf.* **2022**, *58*, 103042. [CrossRef]
2. Aldalur, E.; Suárez, A.; Veiga, F. Metal transfer modes for Wire Arc Additive Manufacturing Al-Mg alloys: Influence of heat input in microstructure and porosity. *J. Mater. Process. Technol.* **2021**, *297*, 117271. [CrossRef]
3. Sinha, A.K.; Pramanik, S.; Yagati, K.P. Research progress in arc based additive manufacturing of aluminium alloys—A review. *Measurement* **2022**, *200*, 111672. [CrossRef]
4. Wu, B.; Pan, Z.; Ding, D.; Cuiuri, D.; Li, H.; Fei, Z. The effects of forced interpass cooling on the material properties of wire arc additively manufactured Ti6Al4V alloy. *J. Mater. Process. Technol.* **2018**, *258*, 97–105. [CrossRef]
5. Wang, X.; Zhu, C.; Xie, H.; Li, H.; Liu, F.; Ren, W.; Tan, F. A high-efficiency forming method of large-scale aluminum alloy components using fused deposition additive manufacturing based on multi-energy fields. *J. Manuf. Process.* **2021**, *62*, 704–710. [CrossRef]
6. Singh, S.R.; Khanna, P. Wire arc additive manufacturing (WAAM): A new process to shape engineering materials. *Mater. Today Proc.* **2021**, *44*, 118–128. [CrossRef]

7. Lee, J.H.; Lee, C.M.; Kim, D.H. Repair of damaged parts using wire arc additive manufacturing in machine tools. *J. Mater. Res. Technol.* **2022**, *16*, 13–24. [CrossRef]
8. Liu, Y.; Liu, Z.; Zhou, G.; He, C.; Zhang, J. Microstructures and properties of Al-Mg alloys manufactured by WAAM-CMT. *Materials* **2022**, *15*, 5460. [CrossRef]
9. Li, K.; Yang, T.; Hou, X.; Ji, C.; Zhu, L.; Li, B.; Cao, Y.; Zhao, L.; Ma, C.; Tian, Z. Microstructure evolution and mechanical properties of Al–Cu–Mn–Cd alloy fabricated by CMT-wire arc additive manufacturing. *Mater. Sci. Eng. A* **2024**, *898*, 146395. [CrossRef]
10. Zhou, S.; Xie, H.; Ni, J.; Yang, G.; Qin, L.; Guo, X. Metal transfer behavior during CMT-based Wire Arc Additive Manufacturing of Ti-6Al-4V alloy. *J. Manuf. Process.* **2022**, *82*, 159–173. [CrossRef]
11. Wu, B.; Pan, Z.; Ding, D.; Cuiuri, D.; Li, H.; Xu, J.; Norrish, J. A review of the wire arc additive manufacturing of metals: Properties, defects and quality improvement. *J. Manuf. Process.* **2018**, *35*, 127–139. [CrossRef]
12. Jiang, P.F.; Li, X.R.; Zong, X.M.; Wang, X.B.; Chen, Z.K.; Yang, H.X.; Liu, C.Z.; Gao, N.K.; Zhang, Z.H. Multi-wire arc additive manufacturing of Ti basic heterogeneous alloy: Effect of deposition current on the microstructure, mechanical property and corrosion-resistance. *J. Alloys Compd.* **2022**, *920*, 166056. [CrossRef]
13. Li, Y.; Su, C.; Zhu, J. Comprehensive review of wire arc additive manufacturing: Hardware system, physical process, monitoring, property characterization, application and future prospects. *Results Eng.* **2022**, *13*, 100330. [CrossRef]
14. Jafari, D.; Vaneker, T.H.J.; Gibson, I. Wire and arc additive manufacturing: Opportunities and challenges to control the quality and accuracy of manufactured parts. *Mater. Des.* **2021**, *202*, 109471. [CrossRef]
15. Gisario, A.; Kazarian, M.; Martina, F.; Mehrpouya, M. Metal additive manufacturing in the commercial aviation industry: A review. *J. Manuf. Syst.* **2019**, *53*, 124–149. [CrossRef]
16. Abouel Nour, Y.; Gupta, N. In-situ monitoring of sub-surface and internal defects in additive manufacturing: A review. *Mater. Des.* **2022**, *222*, 111063. [CrossRef]
17. Hauser, T.; Da Silva, A.; Reisch, R.T.; Volpp, J.; Kamps, T.; Kaplan, A.F. Fluctuation effects in Wire Arc Additive Manufacturing of aluminium analysed by high-speed imaging. *J. Manuf. Process.* **2020**, *56*, 1088–1098. [CrossRef]
18. Zhu, L.; Luo, Y.; Han, J.; Zhang, C.; Xu, J.; Chen, D. Energy characteristics of droplet transfer in wire-arc additive manufacturing based on the analysis of arc signals. *Measurement* **2019**, *134*, 804–813. [CrossRef]
19. Ke, W.C.; Oliveira, J.P.; Cong, B.Q.; Ao, S.S.; Qi, Z.W.; Peng, B.; Zeng, Z. Multi-layer deposition mechanism in ultra high-frequency pulsed wire arc additive manufacturing (WAAM) of NiTi shape memory alloys. *Addit. Manuf.* **2022**, *50*, 102513. [CrossRef]
20. Doumanidis, C.; Kwak, Y.M. Multivariable adaptive control of the bead profile geometry in gas metal arc welding with thermal scanning. *Int. J. Press. Vessel. Pip.* **2002**, *79*, 251–262. [CrossRef]
21. Choudhury, S.S.; Marya, S.K.; Amirthalingam, M. Improving arc stability during wire arc additive manufacturing of thin-walled titanium components. *J. Manuf. Process.* **2021**, *66*, 53–69. [CrossRef]
22. Shi, X.; Cai, C.; Bao, P.; Li, Z. Influence of ship-based vibration on characteristics of arc and droplet and morphology in wire arc additive manufacturing. *Chin. J. Mech. Eng. Addit. Manuf. Front.* **2023**, *2*, 100067. [CrossRef]
23. Guo, Y.; Zhang, Y.; Pan, Z.; Zhou, W. Recent progress of sensing and machine learning technologies for process monitoring and defects detection in wire arc additive manufacturing. *J. Manuf. Process.* **2024**, *125*, 489–511. [CrossRef]
24. Fu, R.; Tang, S.; Lu, J.; Cui, Y.; Li, Z.; Zhang, H.; Xu, T.; Chen, Z.; Liu, C. Hot-wire arc additive manufacturing of aluminum alloy with reduced porosity and high deposition rate. *Mater. Des.* **2021**, *199*, 109370. [CrossRef]
25. Zhang, C.; Gao, M.; Chen, C.; Zeng, X. Spectral diagnosis of wire arc additive manufacturing of Al alloys. *Addit. Manuf.* **2019**, *30*, 100869. [CrossRef]

materials

MDPI

Article

A Multi-Layer Multi-Pass Weld Bead Cross-Section Morphology Extraction Method Based on Row–Column Grayscale Segmentation

Ting Lei [1,2], Shixiang Gong [1,2] and Chaoqun Wu [1,2,*]

1 School of Mechanical and Electrical Engineering, Wuhan University of Technology, Wuhan 430070, China; leiting0621@whut.edu.cn (T.L.); gsx1371470161@whut.edu.cn (S.G.)
2 Hubei Provincial Engineering Research Center of Robotics and Intelligent Manufacturing, Wuhan 430070, China
* Correspondence: chaoqunwu@163.com

Abstract: In the field of welding detection, weld bead cross-section morphology serves as a crucial indicator for analyzing welding quality. However, the extraction of weld bead cross-section morphology often relies on manual extraction based on human expertise, which can be limited in consistency and operational efficiency. To address this issue, this paper proposes a multi-layer multi-pass weld bead cross-section morphology extraction method based on row–column grayscale segmentation. The weld bead cross-section morphology image is pre-processed and then segmented into rows and columns based on the average gray value of the image. In order to extract the feature of multi-layer multi-pass weld feature images, an outline showing the binarization threshold is selected for each segmented image (ESI). Then, the weld contour of ESI is extracted before image fusion and morphological processing. Finally, the weld feature parameters (circumference, area, etc.) are extracted from the obtained weld feature image. The results indicate that the relative errors in circumference and area are within 10%, while the discrepancies in maximum weld seam width and maximum weld seam height can be close to the true value. The quality assessment falls within a reasonable range, the average value of SSIM is above 0.9 and the average value of PSNR is above 60 on average. The results demonstrate that this method is feasible for extracting the general contour features of multi-layer multi-pass weld bead cross-section morphology images, providing a basis for further detailed analysis and improvement in welding quality assessment.

Keywords: weld bead morphology; image identification; multi-layer multi-pass; feature extraction method

1. Introduction

In the current context of continuously increasing demands for product quality and safety, welding technology, as a well-established material thermal processing technique, has consistently been applied across various manufacturing industries such as vehicles, ships, bridges, and aerospace. Simultaneously, in products featuring welded joint structures, the quality of welds plays a decisive role in determining the final product's quality and safety performance. For instance, in the process of battery production, there is a critical demand for high-quality welding inspection [1].

The quality of weld seam formation is a crucial criterion for evaluating weld quality. A well-formed weld bead should exhibit a surface with uniform, delicate, and aesthetically pleasing ripples. It should also possess correct geometric shapes, appropriate weld seam height, smooth transition between the weld seam and the base material and be devoid of important defects such as undercut and lack of fusion. Post-weld inspection of the weld appearance is an effective method to ensure and enhance the quality of weld seam formation.

Recently, welding methods have become increasingly diverse, resulting in varying characteristics of weld seams. For instance, welding dissimilar materials can lead to asymmetrical features [2]. Laser welding offers high operability but tends to exhibit keyhole features [3]. Multi-layer and multi-pass welding can also result in certain angular distortions [4]. Different steels require different welding methods, and the weldability of steel decreases as the carbon content increases [5]. When the welding speed is increased, there will be some new problems such as undercut and hump, which differ from those of traditional welding speeds [6].

Therefore, when conducting quality inspection of welds, the cross-sectional morphology of the weld will also be an important indicator for evaluating the weld. Currently, the traditional method of appearance inspection involves manual observation by inspectors and simple measurements using inspection tools. This approach often results in substantial workload and lacks rigorous scientific conclusions due to variations in experience and qualifications among inspectors. When faced with hundreds or thousands of inspections, the process is inefficient and has limitations in consistency. Moreover, it primarily provides qualitative detection of defects, lacking a quantitative assessment. In the contemporary context of highly automated welding production, there is a need for a feasible visual automated system for detecting weld morphology. However, traditional ultrasound imaging systems incur high costs in image acquisition [7].

To overcome the reliance on traditional quality inspection methods, which are dependent on experience, have low accuracy, and require long identification times [8], in the contemporary context of highly automated welding production, there is a need for an automated visual inspection system.

Siva Shanmugam et al. [9] used finite element research to carry out three-dimensional finite element modeling of weld morphology and predict parameters. However, such methods require high fitting accuracy of formulas and cannot fully express small feature gaps of welds. Zhao et al. [10] utilized computational fluid dynamics (CFD) to predict weld bead cross-sectional features, but this approach is limited to single-pass welds and cannot accurately depict actual weld cross-section morphology.

Wójcicka et al. [11] identified weld seams by image histogram equalization, filtering, and adaptive binarization and segmentation of the image, but it has more limitations. Ai et al. [12] used segmentation of weld seams using a seed region growing method with automatic selection of initial seeds, which can better extract the weld morphology, but the examined weld features are not complete enough. Zhang et al. [13] adopted adaptive image fusion technology to detect internal features of weld seams, thereby enabling the adaptive extension of the dynamic range of vision detection systems at a relatively low cost. Ting Lei et al. [14] proposed a dynamic threshold adjustment method, which can combine the quick positioning of the Otsu method and the weight balance of the average method which plays a proactive role in monitoring the morphology of small holes in laser welding.

Sun Jiahao et al. [15] employed a neural network to predict the weld bead cross-section morphology of 316L stainless steel laser welding, describing the weld seam contour using Hermite interpolation and least squares fitting. However, the coordinates of the weld seam contour points were obtained from measurement software MATLAB R2019a, introducing inaccuracies in the coordinates. The detection of welding seam images was conducted by Alaknanda et al. [16] using a morphological watershed algorithm, which addressed the issue of over-segmentation in single-layer watershed segmentation. However, the accuracy of extracting the morphology contour of the welding seam was still insufficient.

Therefore, this paper, with the identification and detection of weld seam cross-section morphology as the starting point, proposes a method for processing weld seam cross-section morphology images based on row–column grayscale segmentation. This method enables the automated processing and contour extraction of collected weld cross-section morphology images, obtaining weld characteristic parameters. This represents a new application of image processing technology in the field of weld identification, potentially advancing more precise contour extraction of welds and the prediction of welding process parameters.

2. Materials and Methods

2.1. Materials Preparation

In this study, the raw materials used were 240 mm × 80 mm × 20 mm thick Q235 steel (Wuhan Hezeshun Steel Co., Ltd., Wuhan, China) plates with a 45° bevel and a 4 mm root face, shown in Figure 1a. The welding process employed was Gas Metal Arc Welding (GMAW), utilizing H08MnSiA welding wire(Nangong Jingkun Welding Materials Sales Co., Ltd., Xingtai, China) as the filler material. A V-groove multi-pass welding technique was applied, consisting of three layers and six passes, arranged in an inverted triangular pattern within the V-groove. The welding process began with the first-layer first pass, measuring 240 mm in length. Each subsequent pass decreased in length by 40 mm. Specifically, the second pass and third pass of the second layer measured 200 mm and 160 mm, respectively, while the sixth pass of the third layer measured 40 mm.

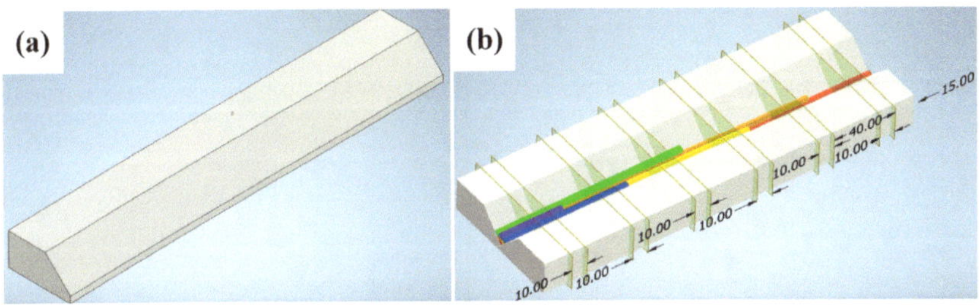

Figure 1. Schematic diagram of material preparation. (**a**) Base material. (**b**) Schematic of welding and EDM.

The welded material, with multi-pass weld beads, was cut using wire electrical discharge machining (EDM). The initial cutting position was 15 mm from the edge of the base material, with additional cuts made at 40 mm intervals, resulting in six cuts to obtain cross-sections from the first pass of the first layer to the sixth pass of the third layer. The cutting distance for the EDM was 10 mm, shown in Figure 1b.

After cutting, the samples were successively polished with sandpapers of 800, 1200, 1500, and 2000 grit. Upon completion of the sanding process, the samples were polished using a polishing machine(Changsha Miqi instrument equipment Co., Ltd., Changsha, China) equipped with a silk velvet polishing cloth. The polishing cloth was sprayed with W2.5 diamond suspension for polishing.

The samples were then etched with aqua regia for 7 s and immediately rinsed with water. Aqua regia was prepared by mixing hydrochloric acid (37% by mass) and nitric acid (65% by mass) in a 3:1 volume ratio, with the etching process conducted at room temperature. The prepared samples were observed under a Leica high-power microscope (Leica, Wetzlar, Germany) at a magnification of 16×.

2.2. Features of Weld Bead Cross-Section Morphology

The weld bead cross-section morphology refers to the cross-section of the weld joint after completion. This process involves dissecting the welded joint, followed by using etching techniques to enhance the clarity of the weld bead's structure, making it easier to observe and analyze the internal features of the weld bead [17]. Weld bead morphology is observed visually, primarily comprising geometric parameters of the appearance such as weld seam area (WSA), weld seam circumference (WSC), maximum weld seam height (MWSH) and maximum weld seam width (MWSW) as illustrated in Figure 2. Due to processes such as high-temperature heat transfer, melting and solidification, and metal phase transformation involved in welding [9,18], there exists a highly nonlinear relationship between welding process parameters and weld bead profiles. The characteristics of weld bead cross-section morphology are related to many factors; for instance, significant differences exist in weld bead cross-section morphology under different welding speed conditions [19]. And the other factors include the welding current and voltage, different welding methods, weld height, material properties, etc.

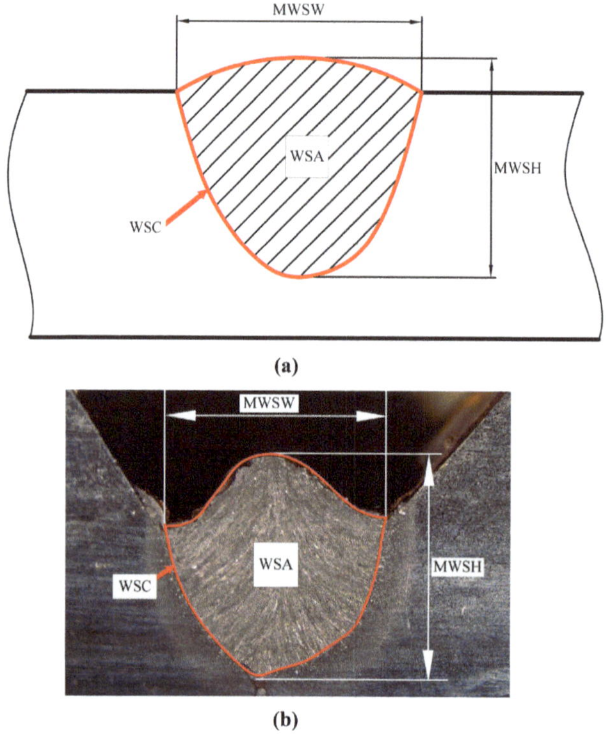

Figure 2. Schematic diagram of weld bead cross-section morphology. (**a**) Schematic diagram. (**b**) Physical schematic diagram.

2.3. *Grayscale Segmentation Method for Weld Bead Sectional Images*

Shown in Figure 3, the technical approach of this paper involves establishing a welding seam detection platform, followed by obtaining the original images of the welding bead sections. These images are then grayscale-converted and subjected to row–column grayscale segmentation. Specifically, the segmentation method calculates the average grayscale value for each row and column in the grayscale image. The total grayscale value of the image is computed, and then, through iteration, the point of segmentation along a row/column is found where the total grayscale values of the upper/lower or left/right halves are equal. This point serves as the boundary for segmentation. After segmentation, the ESI is separately subjected to binarization and median filtering to retain only the main weld bead cross-section morphology. The segmented images are then recombined, and morphological processing is applied to remove irregularities along the edges. Finally, feature extraction is performed on the processed images to obtain welding bead characteristic data.

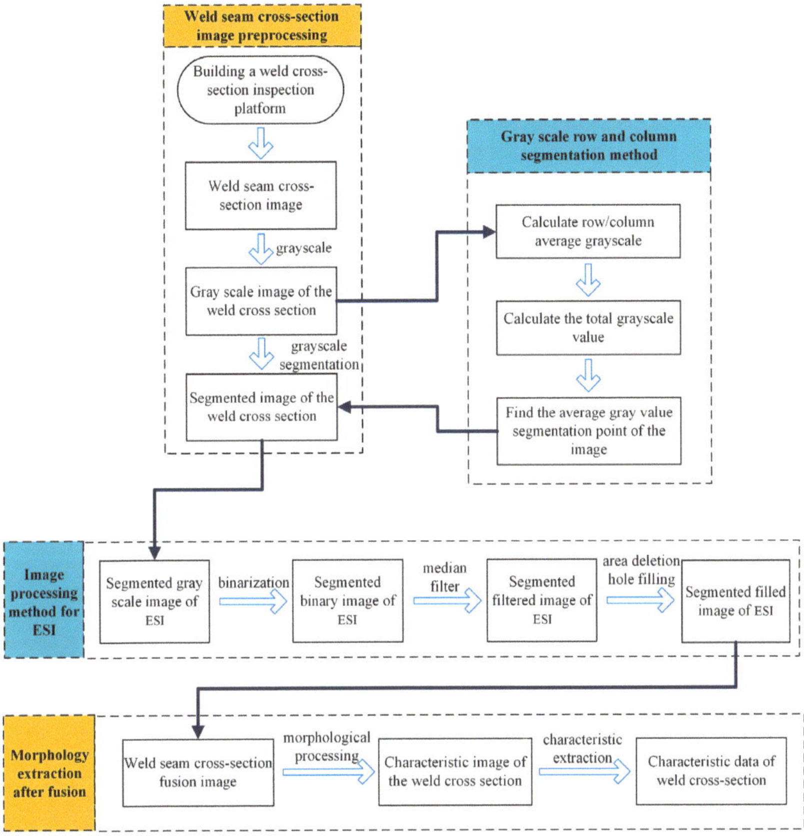

Figure 3. Flowchart of the grayscale segmentation method.

2.4. Gray Scale Analysis of a Weld Bead Cross-Section Image

Images of regions with similar grain structures often exhibit comparable grayscale characteristics. The grayscale values corresponding to the base material and the weld area typically fall within different ranges. Therefore, we can utilize pixel-based grayscale values to process images, extracting the grayscale range corresponding to the weld area and separating it from the base material to obtain the weld bead's contour.

The issue of uneven grayscale distribution in weld bead images presents a significant challenge in the field of welding image processing. This uneven grayscale distribution often stems from the instability of the light source at the welding site and variations in illumination conditions, thereby directly impacting the visual characteristics of weld bead images. This phenomenon presents formidable challenges for traditional image processing algorithms in the recognition of overall weld beads.

When directly processing the weld bead cross-section morphology of the third-layer sixth pass, differences in grayscale values are observed in the images, shown in Figure 4. Consequently, this leads to uneven phenomena during the binarization process.

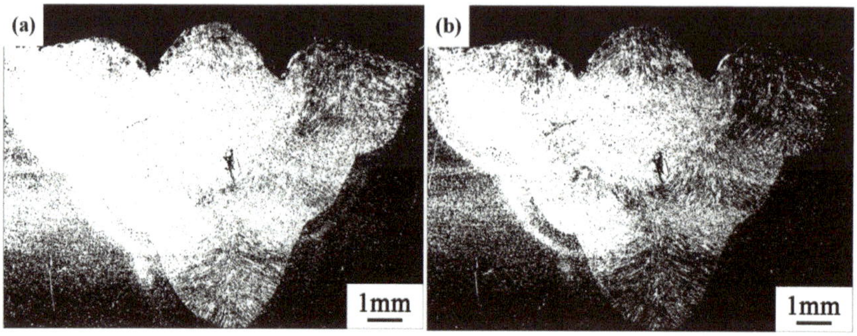

Figure 4. The uneven phenomena during the binarization process. (**a**) Threshold of 75. (**b**) Threshold of 90.

As shown in Figure 5, after analyzing the 3D grayscale image of the weld, a noticeable gradient in grayscale values is observed. Specialized processing is required based on these characteristics to address the uneven grayscale distribution in certain regions, aiming to achieve improved results.

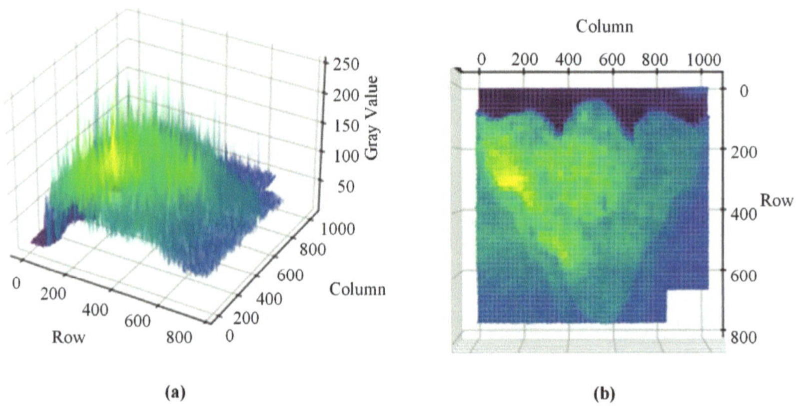

Figure 5. The 3D grayscale image of the weld bead. (**a**) Axial view. (**b**) Top view.

An innovative row–column-based grayscale image segmentation method has been employed to address the challenge of uneven grayscale distribution in weld bead images. Through this approach, local optimization of the weld bead at a microscopic level is achieved in the image. Partitioning along the rows and columns of the image enables the intricate capture of local features in the weld bead image, allowing selective adjustment of areas with uneven brightness. This approach significantly enhances the accuracy of subsequent processing.

2.5. Principle of Row and Column Average Gray Value Segmentation

When performing segmentation on grayscale images, there are two approaches: one is based on the peaks of the row–column grayscale image, and the other is based on the average values of the left and right sides of the row–column grayscale image. The advantage of using peak values for segmentation lies in its ability to sensitively capture prominent edges or features in the image, as peaks often correspond to significant changes or boundaries of the target in the image. By identifying peaks in the grayscale image, the image can be segmented into different regions, where the grayscale values change more prominently within each region. However, this method may exhibit excessive sensitivity when dealing with regions where the brightness gradient is relatively smooth. It can be prone to noise interference, leading to unstable segmentation results.

Compared to the peak-based method, the segmentation approach based on average values is more stable and robust. This method calculates the average grayscale value for each row or column and determines the segmentation point based on these averages. This approach is relatively smooth, making it better suited for regions with gradual changes in grayscale values. It is also less susceptible to noise interference. Additionally, the calculation of the average values is relatively simple, leading to higher computational efficiency.

According to the principle of segmentation based on column-wise average grayscale values, the formula is as follows:

$$H = \begin{bmatrix} h_{11} & \cdots & h_{1n} \\ \vdots & \ddots & \vdots \\ h_{m1} & \cdots & h_{mn} \end{bmatrix} \tag{1}$$

$$N_j = \frac{\sum_{i=1}^{m} h_{ij}}{m} \tag{2}$$

$$N = \begin{bmatrix} N_1 & \cdots & N_j & \cdots & N_n \end{bmatrix} \tag{3}$$

$$S_{N_j} = \sum_{k=1}^{j} N_k \tag{4}$$

$$S_{N_n} = \sum_{k=1}^{n} N_k \tag{5}$$

$$S_{N_j} \geq \frac{1}{2} S_{N_n} \tag{6}$$

where h_{ij} represents the grayscale value of the individual pixel at the i-th row and j-th column, H denotes the set of grayscale values for an image with length m and width n, N_j is the average grayscale value for the j-th column, N represents the set of average grayscale values for each column, and S_{N_j} is the sum of grayscale values from the 1-st column to the j-th column. In the process of determining the segmentation point, a traversal of the j values is conducted. The segmentation column, denoted as $\bar{j}$, is identified as the first j value that satisfies the inequality S_{N_j} which is greater than or equal to the total grayscale value of the image, S_{N_n}.

Similarly, based on the principle of segmentation using row-wise average grayscale values, the formula is as follows:

$$M_i = \frac{\sum_{j=1}^{n} h_{ij}}{n} \tag{7}$$

$$M = \begin{bmatrix} M_1 \\ \vdots \\ M_i \\ \vdots \\ M_m \end{bmatrix} \tag{8}$$

$$S_{M_i} = \sum_{k=1}^{i} M_k \tag{9}$$

$$S_{M_m} = \sum_{k=1}^{m} M_k \tag{10}$$

$$S_{M_i} \geq \frac{1}{2} S_{M_m} \tag{11}$$

In this case, in the process of determining the segmentation point, a traversal of the i values is conducted. The segmentation row, denoted as $\bar{i}$, is identified as the first i value that satisfies the inequality S_{M_i} which is greater than or equal to the total grayscale value of the image, S_{M_m}.

2.6. Weld Bead Cross-Section Image Processing Methods

2.6.1. Basic Image Processing Methods for Weld Bead Cross-Sections

During the process of capturing weld bead cross-section morphology images, environmental factors and the selection of equipment and facilities may introduce defects in the acquired images. Furthermore, the errors in these images can accumulate and propagate. Therefore, preprocessing of the images is essential to filter out irrelevant noise and interference while retaining the essential information. Dealing with color images can increase the complexity of algorithms. Hence, grayscale images are often used before image processing, or the color images are converted to grayscale. The original weld bead cross-section morphology images inherently possess complexity due to their internal crystal structure and the influence of the cutting profile. To extract contours from these images, it is necessary to process and fill the internal regions. Therefore, thresholding, median filtering, region deletion, and hole-filling operations are applied to achieve effective preprocessing.

1. **Binarize the image after row and column grayscale segmentation**

Fixed thresholding algorithms require manual thresholding, although this process can be time consuming. However, when dealing with illustrations with some complexity and small differences between the noise and the target pattern, a more accurate processing of the pattern can be achieved by manually adjusting the fixed threshold for binarization operations. This fine manual adjustment process helps to cope with subtle differences in complex images and suppresses noise while preserving the target features, thus improving the accuracy of pattern binarization.

2. **Binary image median filtering**

Median filtering is a nonlinear filtering method that is effective in filtering isolated point type noise. The basic principle is to set the pixel value at the anchor point to the median of the grayscale values in the core window centered on the anchor point [20].

3. **Obtain the maximum contour of the filtered image**

The shape of the weld cut surface has many irregular crystal features. Even after the completion of the binarization and median filtering process, there will still be many small regions with scattered distribution, which will bring a series of subsequent detection of many difficulties; thus, the image has been completed to remove most of the noise for the contour detection, and it will be less than the set threshold for the removal of the region which will leave only the largest contour.

4. **Fill the contour image to obtain the weld profile image**

The resulting maximum contour is found by drawing it in a new image and filling in the inner regions to obtain a complete, edge-continuous and noise-free contour of the weld bead.

2.6.2. Edge Detection and Feature Extraction of Weld Bead Cross-Section Image

In the former step, the region containing the weld bead cross-section morphology was identified, and its contour was subsequently detected. Consequently, edge detection and feature extraction can be conducted on this contour to facilitate quantitative analysis of the weld bead cross-section morphology. The pixel values of this largest contour are analyzed to identify the leftmost and rightmost pixel points, thereby determining its MWSW, along with the topmost and bottommost pixel points to ascertain its MWSH.

3. Results

3.1. Preparation for Sectional Image Row–Column Gray Segmentation

3.1.1. The Original Weld Bead Image Data

Before conducting threshold analysis on the segmented image, the contour area of the original image is manually selected and contour extraction is performed, as shown in Figure 6. The collected data includes the following:

WSC: 3289, WSA: 404,529.5, MWSH: 728, and MWSW: 1024.

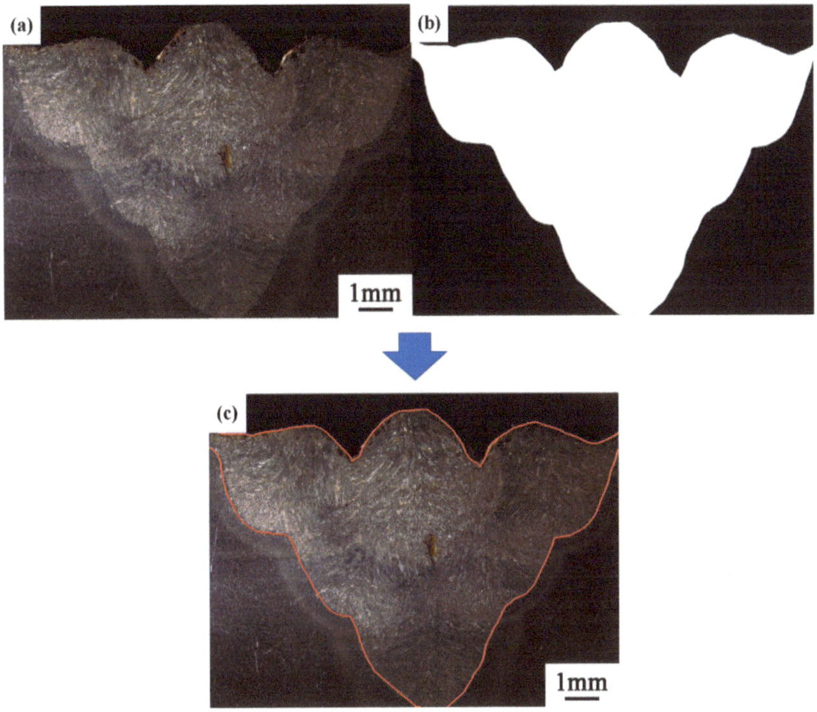

Figure 6. The sample of the third-layer sixth-pass weld bead cross-section image. (**a**) The original image. (**b**) The manual contour. (**c**) The original image overlaid with manual contour.

3.1.2. Sectional Image Gray Segmentation Analysis

After grayscale conversion of a sample image depicting the third-layer sixth-pass weld bead cross-section morphology, the pixel values were summed and averaged for each row or column. The resulting curves of the row-wise and column-wise average grayscale values are illustrated in Figure 7.

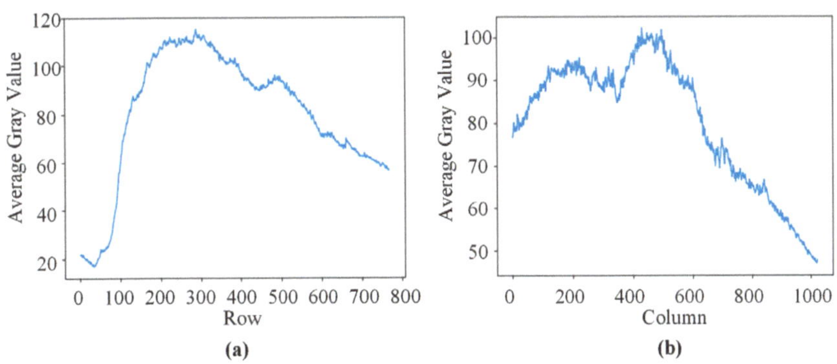

Figure 7. Average grayscale histogram. (**a**) The average gray value of each row. (**b**) The average gray value of each column.

From the average grayscale plot, it is evident that the main feature of the image is brightness in the upper region and darkness in the lower region, with the left side appearing brighter than the right side. It is worth noting that, for the subsequent image, the brightness difference between the left and right sides is even more pronounced.

3.2. Gray Average Binarized Image

3.2.1. Binarized Method Threshold Analysis

Following image segmentation, binarization was conducted on the left and right sides of the image, with particular emphasis on delineating the threshold ranges. Initially, a threshold of 90 was set as a baseline for processing the left half of the image. Despite subsequent image processing, certain edge regions remained inadequately identified. Consequently, the threshold was incremented; however, raising it to 100 resulted in partial omission of the weld bead cross-section morphology in the upper region. When the threshold value is varied from 90 to 100, the threshold value 95 was identified as the optimal binarization threshold, yielding satisfactory binarization efficacy for the left half of the image. The difference of these thresholds is shown in Figure 8.

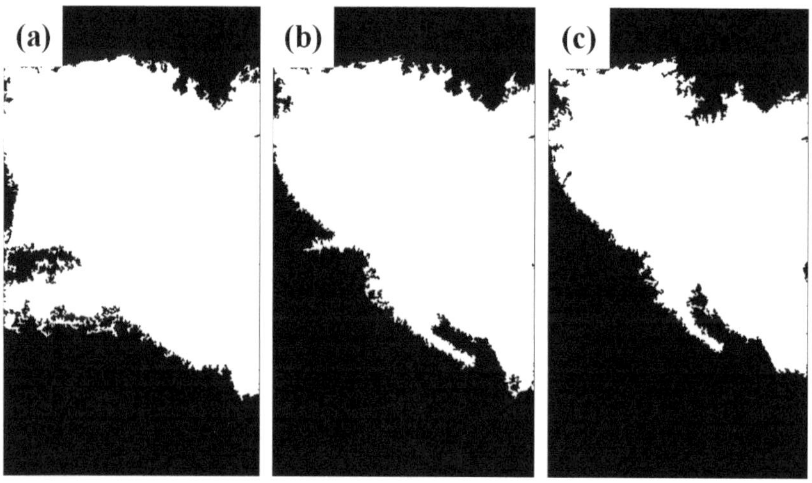

Figure 8. The left half binarized images with different thresholds: (**a**) 90, (**b**) 95, and (**c**) 100.

When addressing the right half of the image, an initial threshold of 90 was employed as a baseline, revealing numerous omitted details. Subsequently, lowering the threshold became imperative to enhance image processing outcomes. By systematically reducing the threshold in increments of five, the threshold that aligned with optimal image processing requisites is progressively identified. The difference of these thresholds is shown in Figure 9.

When the threshold value was varied from 90 to 65, the threshold value 65 was observed when the weld bead cross-section morphology of the right half of the image was well-displayed. Consequently, this threshold value was selected as the final processing result for the right half.

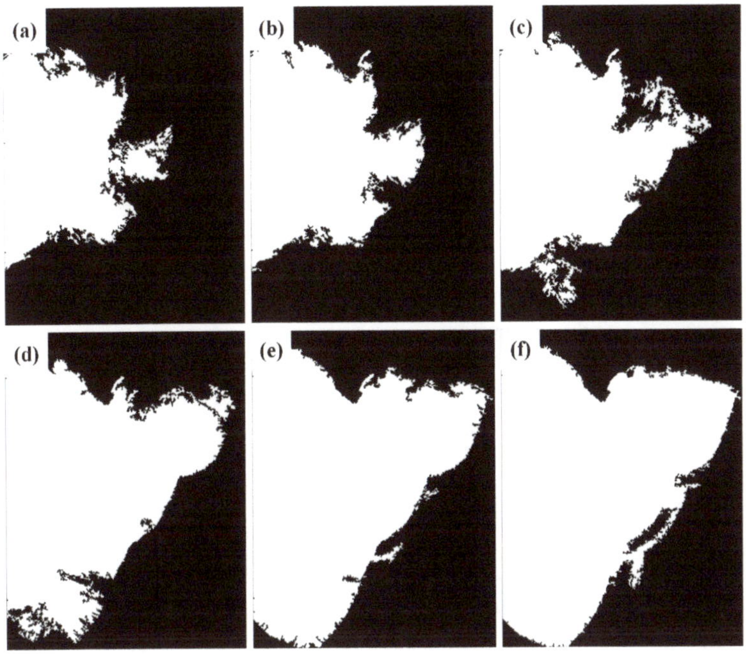

Figure 9. The right half binarized images with different thresholds: (**a**) 90, (**b**) 85, (**c**) 80, (**d**) 75, (**e**) 70, and (**f**) 65.

3.2.2. Fusion of Left and Right Profile Images

The results of processing the left and right sides are fused together, with a comparison made between the individually processed images to determine the one that best displays the weld bead cross-section morphology along with its corresponding threshold. Specifically, the threshold values selected were 95 for the left half image and 65 for the right half image, as shown in Figure 10.

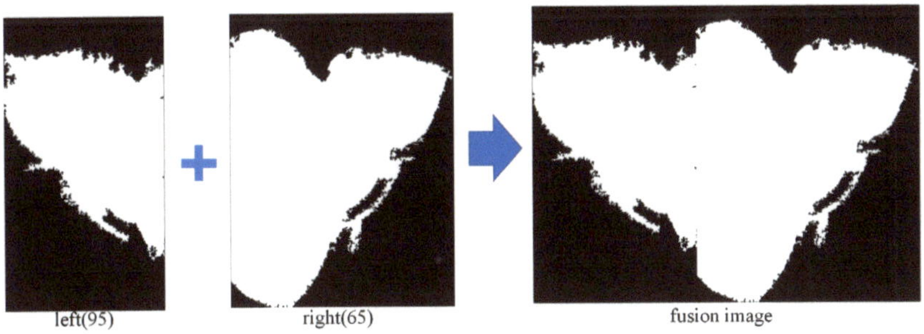

Figure 10. Fusion of left and right profile images.

3.3. Grayscale Averaging Quadripartite Image

3.3.1. Quadripartite Method Threshold Analysis

To address the issue of significant errors observed in the final image compared to the original image in the binary segmentation method, the process is further refined based on average grayscale segmentation. In this step, the image is divided again, this time based on both row and column average grayscale values. The resulting segmented image is then divided into four parts: top-left, top-right, bottom-left, and bottom-right.

After image segmentation, binary processing is applied to the quadripartite images. The key is to determine the threshold range for each quadripartite image. By varying the binary threshold every 5 based on the original image, it is observed that a threshold of 90 maximally preserves the contour of the weld bead cross-section. Consequently, the binary threshold for the top-left quadripartite image is set to 90. Similarly, for the top-right, bottom-left, and bottom-right quadripartite images, the binary thresholds are set to 65, 95, and 70, respectively. The overall threshold sequence is (90, 65, 95, 70). This choice aligns with the characteristics of the image: brighter on top, darker on the bottom, brighter on the left, and darker on the right. The fusion image after processing the four quadripartite images is shown in Figure 11.

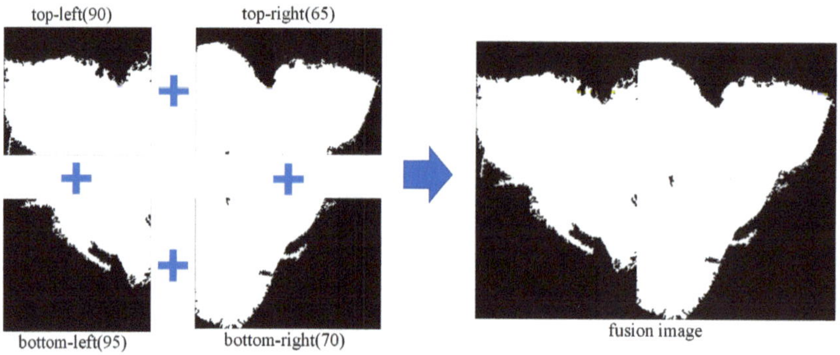

Figure 11. Fusion of the quadripartite profile images.

3.3.2. Morphological Threshold Analysis

After merging the results of processing the four quadripartite images, further morphological processing is required to extract better contours of the image. The obtained morphology contours also depend on the morphological threshold. Here, three threshold values were experimented with, (15, 15), (25, 25), and (35, 35), as shown in Figure 12. The extracted contour data for ESI are presented in Table 1.

Table 1. The contour data of the morphological images with different thresholds.

Threshold	Original Figure	(15, 15)	(25, 25)	(35, 35)
WSC (pixel)	3289	4169	3787	3604
WSA ($pixel^2$)	404,529.5	422,026.5	428,483.0	432,633.5
MWSH (pixel)	728	726	726	726
MWSW (pixel)	1024	1024	1024	1024

After comparing the data, it was observed that as the threshold value increases, the circumference gets closer to the true value, but the area becomes larger, deviating from the true value. However, the difference of MWSH and difference of MWSW remain consistent and are not affected. The relative errors of WSA are 26.8%, 15.1%, and 9.6%, respectively, while the relative errors of WSC are 4.3%, 5.9%, and 6.9%, respectively. The difference of MWSH and difference of MWSW show minimal errors.

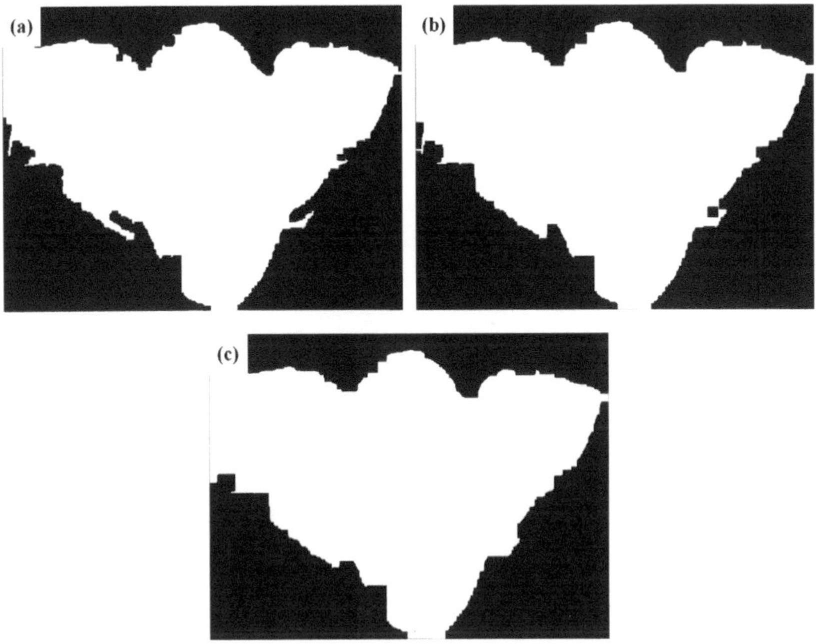

Figure 12. The morphological images with different thresholds: (**a**) (15, 15), (**b**) (25, 25), and (**c**) (35, 35).

4. Discussion

4.1. Repeatability Tests

4.1.1. Repeatability Tests of the Binary Segmentation Method for Multi-Layer Multi-Pass Weld Bead Sectional Images

After conducting inspections on the third-layer sixth-pass weld bead images, to validate the method's applicability, repeatability tests were also performed on weld bead images ranging from the first-layer first-pass to the third-layer fifth-pass configurations, as shown in Figures 13–18. Considering the uneven brightness characteristics of each weld bead image, grayscale segmentation was carried out either horizontally or vertically. Subsequently, the binary threshold was gradually adjusted to achieve the closest possible match with the desired morphology contours from the original images.

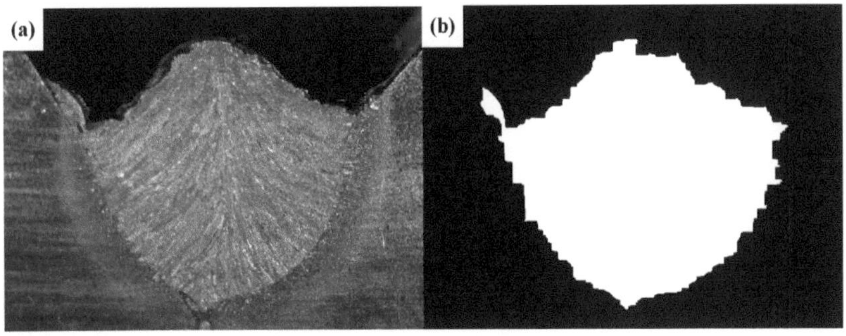

Figure 13. The result of the binary method for the first-layer first-pass image. (**a**) Original image. (**b**) Resulting image.

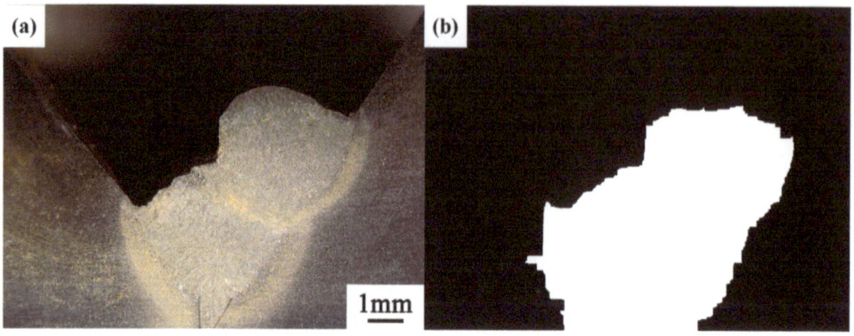

Figure 14. The result of the binary method for the second-layer second-pass image. (**a**) Original image. (**b**) Resulting image.

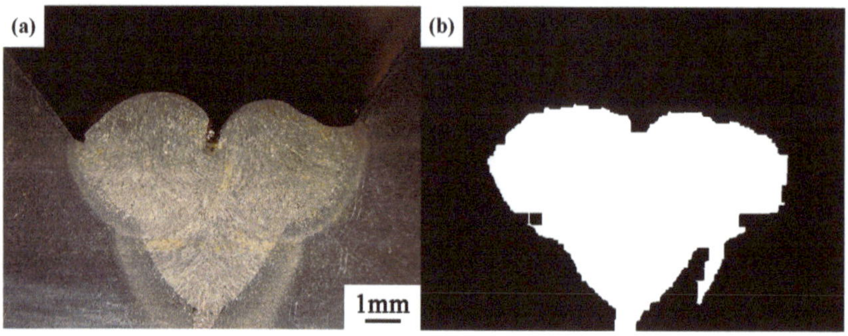

Figure 15. The result of the binary method for the second-layer third-pass image. (**a**) Original image. (**b**) Resulting image.

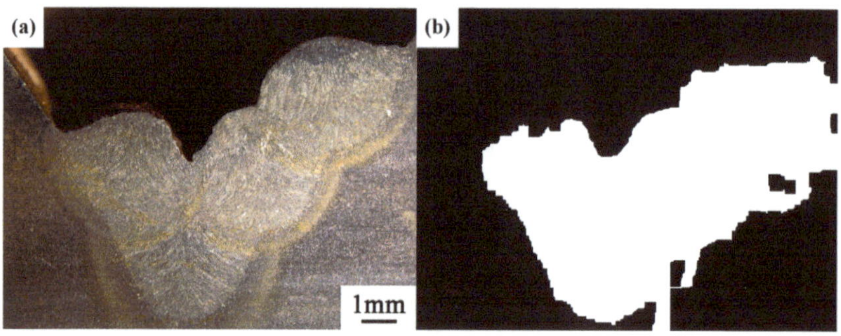

Figure 16. The result of the binary method for the third-layer forth-pass image. (**a**) Original image. (**b**) Resulting image.

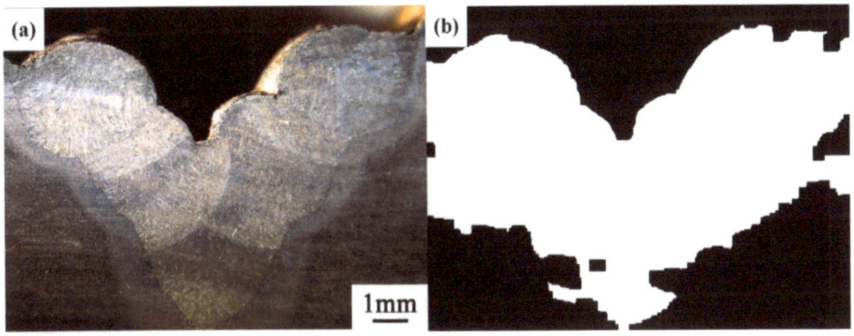

Figure 17. The result of the binary method for the third-layer fifth-pass image. (**a**) Original image. (**b**) Resulting image.

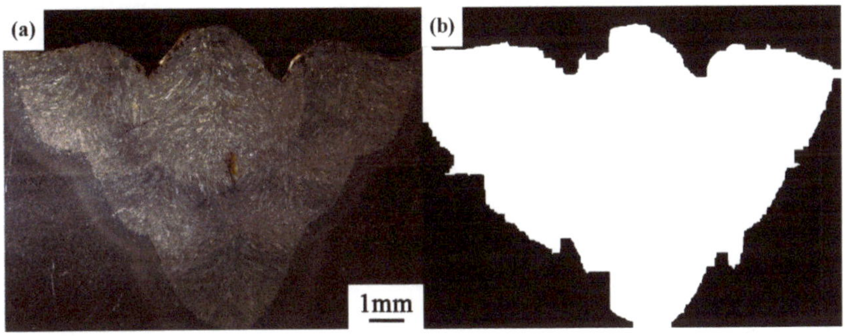

Figure 18. The result of the binary method for the third-layer sixth-pass image. (**a**) Original image. (**b**) Resulting image.

It can be observed that for the weld bead cross-section morphology of the first-layer first-pass configuration, vertical segmentation was employed with binary threshold values of 132 for the upper part and 125 for the lower part. For the weld bead cross-section morphology of the second-layer second-pass configuration, horizontal segmentation was applied with binary threshold values of 150 for the left part and 135 for the right part. The weld bead cross-section morphology of the second-layer third-pass configuration utilized vertical segmentation with binary threshold values of 120 for the upper part and 150 for the lower part. In the case of the weld bead cross-section morphology of the third-layer fourth-pass configuration, horizontal segmentation was employed with binary threshold values of 120 for the left part and 150 for the right part. For the weld bead cross-section morphology of the third-layer fifth-pass configuration, vertical segmentation was applied with binary threshold values of 125 for the upper part and 100 for the lower part. Lastly, the weld bead cross-section morphology of the third-layer sixth-pass configuration utilized horizontal segmentation with binary threshold values of 95 for the left part and 65 for the right part.

It can be noted that different types of weld bead morphologies require distinct segmentation methods and varying binary threshold values. Thus, automated simplification of the entire process remains challenging, indicating the need for further optimization in the future.

4.1.2. Repeatability Tests of Quadripartite Segmentation Method for Multi-Layer Multi-Pass Weld Bead Sectional Images

After validating the method's universality on the third-layer sixth pass of weld bead images, repeatability tests were conducted on weld bead images ranging from the first-layer first pass to the third-layer fifth pass, as shown in Figures 19–24. Each weld bead image's brightness was segmented into four parts. Subsequently, the binary threshold values were gradually adjusted to closely match the required morphological contours in the original images.

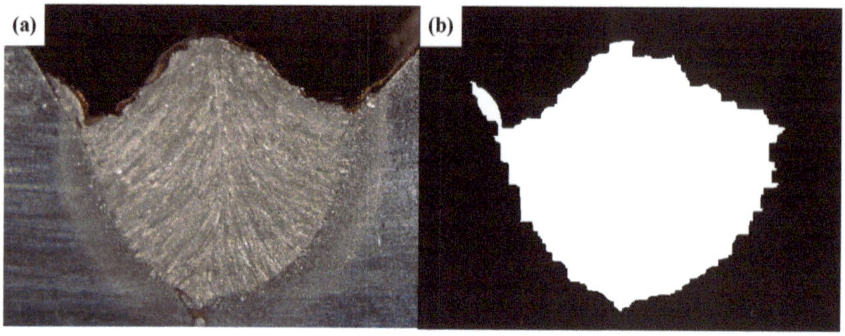

Figure 19. The result of the quadripartite method for the first-layer first-pass image. (**a**) Original image. (**b**) Resulting image.

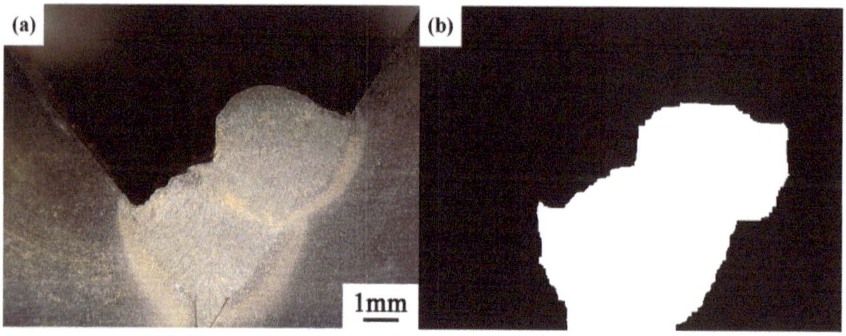

Figure 20. The result of the quadripartite method for the second-layer second-pass image. (**a**) Original image. (**b**) Resulting image.

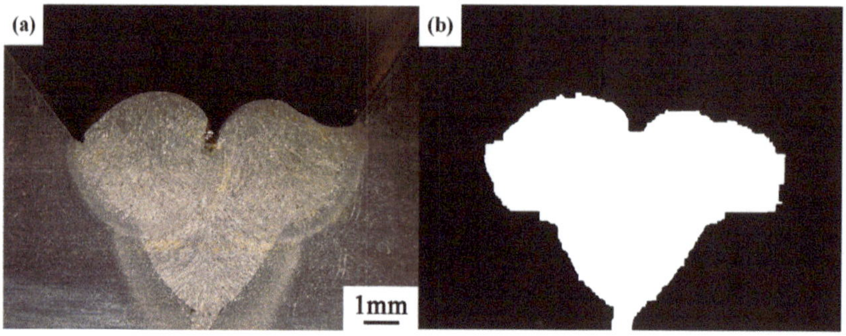

Figure 21. The result of the quadripartite method for the second-layer third-pass image. (**a**) Original image. (**b**) Resulting image.

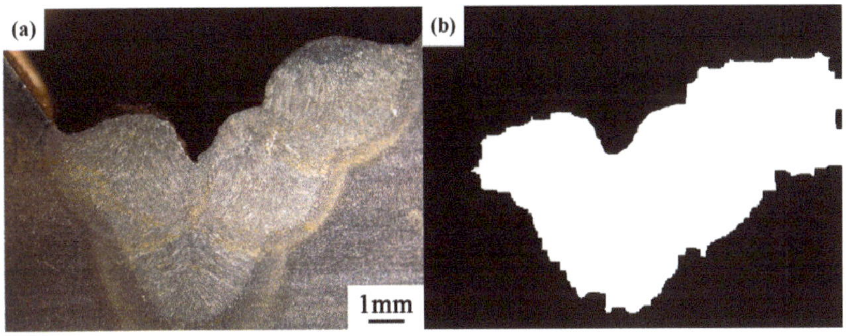

Figure 22. The result of the quadripartite method for the third-layer forth-pass image. (**a**) Original image. (**b**) Resulting image.

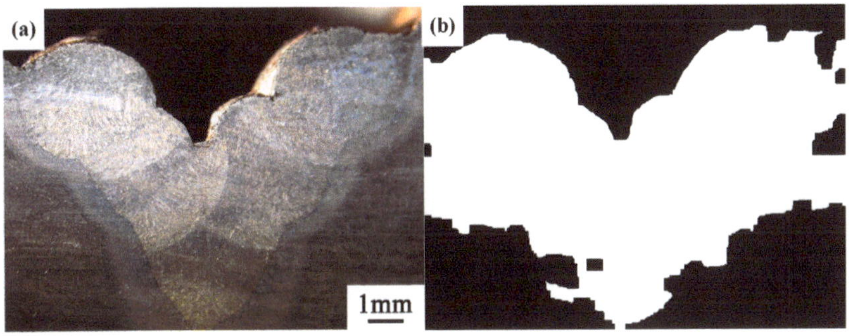

Figure 23. The result of the quadripartite method for the third-layer fifth-pass image. (**a**) Original image. (**b**) Resulting image.

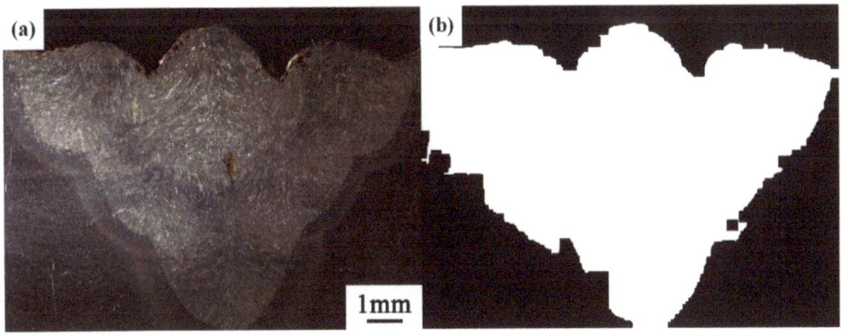

Figure 24. The result of the quadripartite method for the third-layer sixth-pass image. (**a**) Original image. (**b**) Resulting image.

It can be observed that for weld bead morphologies with the first-layer first pass, the binary threshold values are (130, 132, 125, 125). For weld bead morphologies with the second-layer second pass, the binary threshold values are (150, 125, 165, 150). For weld bead morphologies with the second-layer third pass, the binary threshold values are (110, 120, 155, 160). For weld bead morphologies with the third-layer fourth pass, the binary threshold values are (110, 150, 130, 160). For weld bead morphologies with the third-layer fifth pass, the binary threshold values are (120, 130, 100, 95). For weld bead morphologies with the third-layer sixth pass, the binary threshold values are (90, 65, 95, 70).

4.1.3. Comparison of Weld Contour Results

After extracting the weld morphology contours from the first-layer first pass to the third-layer sixth pass, we selected the welds with higher extraction accuracy (from the first-layer first pass to the second-layer third pass) for contour overlay analysis. The actual weld contour was overlaid with the contours extracted in this study on the original images. Figure 25 shows the results obtained using the binary method, while Figure 26 presents the results obtained using the quadripartite method. It can be observed that the contours extracted using the quadripartite method are generally closer to the actual contours and more accurate compared to the binary method.

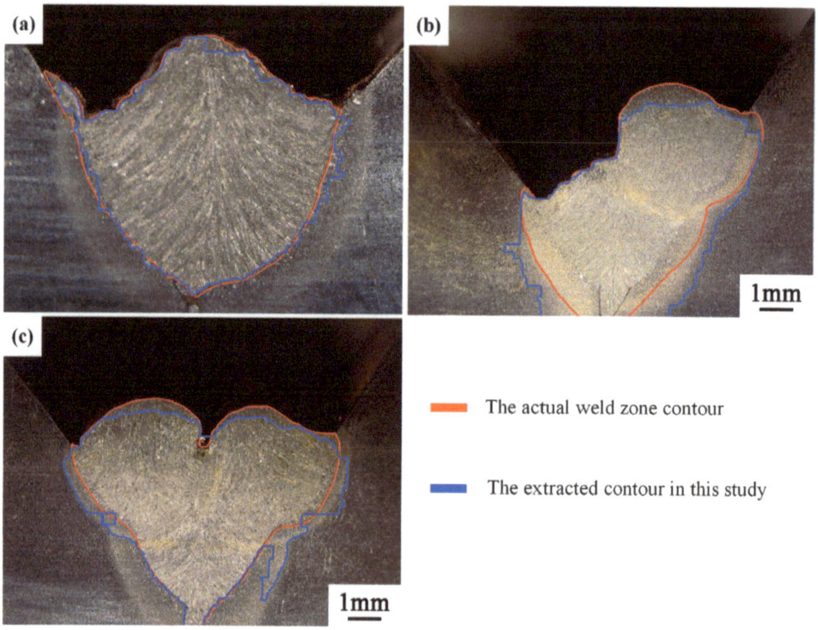

Figure 25. Overlay image of the actual weld zone contour and the contour extracted using the binary method. (**a**) 1st-layer 1st-pass, (**b**) 2nd-layer 2nd-pass, and (**c**) 2nd-layer 3rd-pass.

The method proposed in this study demonstrates a high degree of contour fitting in certain regions, though some areas exhibit deviations that still align with the overall contour trend. Overall, while the method has not yet achieved precise delineation of the weld contour, it is capable of accurately identifying the general morphology of the weld. This is a new exploration in the field of weld detection.

4.2. Analysis and Discussion

4.2.1. Comparison Analysis of Weld Bead Cross-Sectional Feature Data between the Binary Segmentation Method and the Quadripartite Segmentation Method

Following the completion of average grayscale segmentation and the selection of a morphological threshold of 25 for the final image in weld bead cross-section segmentation, the extracted feature data were compared with the original image to compute the errors. Tables 2 and 3 below present the original feature data for multi-layer multi-pass weld bead cross-sections and the feature data after the binary segmentation process.

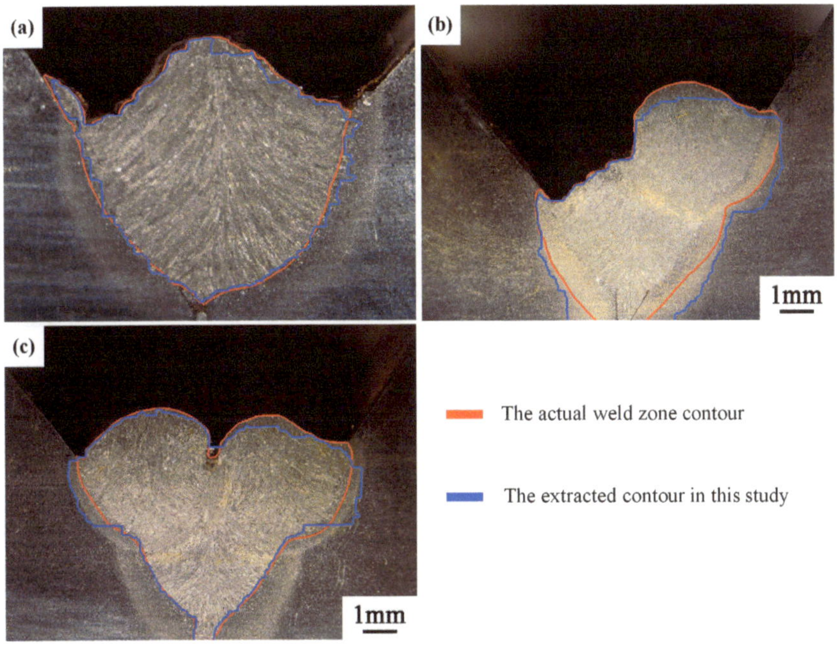

Figure 26. Overlay image of the actual weld zone contour and the contour extracted using the quadripartite method. (**a**) 1st-layer 1st-pass, (**b**) 2nd-layer 2nd-pass, and (**c**) 2nd-layer 3rd-pass.

Table 2. The contour data of the multi-layer multi-pass original images.

Original Image	1st-Layer 1st-Pass	2nd-Layer 2nd-Pass	2nd-Layer 3rd-Pass	3rd-Layer 4th-Pass	3rd-Layer 5th-Pass	3rd-Layer 6th-Pass
WSC (pixel)	2507.2	2027.4	2224.6	2927.0	3419.6	3289.1
WSA (pixel^2)	286,721.5	200,675.0	223,407.0	307,748.0	347,121.5	404,529.5
MWSH (pixel)	651	578	565	701	729	728
MWSW (pixel)	761	609	684	902	1024	1024

Table 3. The contour data of the multi-layer multi-pass images after the binary segmentation process.

Binary Segmentation	1st-Layer 1st-Pass	2nd-Layer 2nd-Pass	2nd-Layer 3rd-Pass	3rd-Layer 4th-Pass	3rd-Layer 5th-Pass	3rd-Layer 6th-Pass
WSC (pixel)	2829.7	2274.7	2763.7	3258.5	4508.1	3739.7
WSA (pixel^2)	277,245.0	216,625.0	230,121.0	309,118.5	440,427.0	434,406.5
MWSH (pixel)	643	535	539	657	721	726
MWSW (pixel)	748	649	728	867	1024	1024

Similar to the binary segmentation of weld cross-sections, a comparative analysis of weld cross-sectional feature data is conducted for the quadripartite segmentation method. The obtained data are in Table 4.

Table 4. The contour data of the multi-layer multi-pass images after the quadripartite segmentation process.

Quadripartite Segmentation	1st-Layer 1st-Pass	2nd-Layer 2nd-Pass	2nd-Layer 3rd-Pass	3rd-Layer 4th-Pass	3rd-Layer 5th-Pass	3rd-Layer 6th-Pass
WSC (pixel)	2884.9	2187.4	2450.4	3215.2	4498.5	3786.6
WSA (pixel^2)	278,755.0	211,807.0	223,177.0	289,484.5	451,563.5	428,483.0
MWSH (pixel)	643	543	562	624	719	726
MWSW (pixel)	767	613	731	908	1024	1024

Due to the observation that errors in the MWSH and MWSW are generally small, our primary focus is on errors in terms of WSC and WSA. The relative error is defined as the ratio of the absolute error of the contour data extracted using the method of this paper to the contour data of the original images. The relative errors of the multi-layer multi-pass images after the binary or quadripartite segmentation process are shown in Tables 5 and 6. We can see a visual comparison in Figure 27.

Table 5. The relative errors of the multi-layer multi-pass images after the binary segmentation process.

Binary Segmentation Error	1st-Layer 1st-Pass	2nd-Layer 2nd-Pass	2nd-Layer 3rd-Pass	3rd-Layer 4th-Pass	3rd-Layer 5th-Pass	3rd-Layer 6th-Pass
WSC	12.86%	12.20%	24.23%	11.32%	31.83%	13.70%
WSA	3.31%	7.95%	3.01%	0.45%	26.88%	7.39%

Table 6. The relative errors of the multi-layer multi-pass images after the quadripartite segmentation process.

Quadripartite Segmentation Error	1st-Layer 1st-Pass	2nd-Layer 2nd-Pass	2nd-Layer 3rd-Pass	3rd-Layer 4th-Pass	3rd-Layer 5th-Pass	3rd-Layer 6th-Pass
WSC	15.07%	7.89%	10.15%	9.85%	31.55%	15.12%
WSA	2.78%	5.55%	0.10%	5.93%	30.09%	5.92%

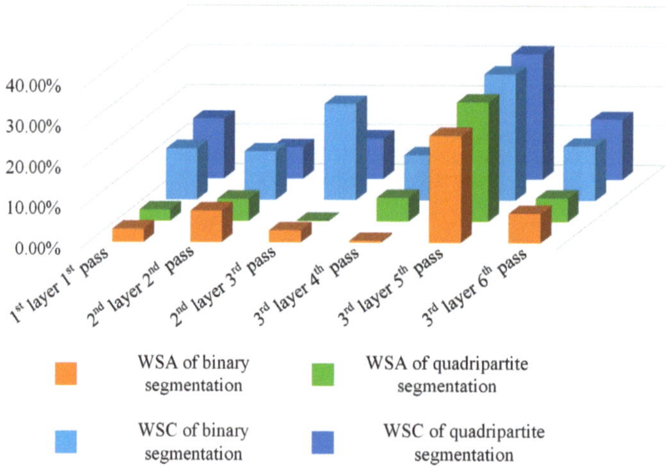

Figure 27. Relative error of the WSC and WSA.

4.2.2. Comparison Analysis of the Quality of Weld Bead Cross-Sectional Images between the Binary Segmentation Method and the Quadripartite Segmentation Method

Furthermore, evaluations were conducted using the Structural Similarity Index (SSIM) and Peak Signal-to-Noise Ratio (PSNR) for the images generated by both binary and quadripartite segmentation methods, in addition to comparing parameters such as circumference and area of the weld bead cross-section morphology. As shown in Figures 28 and 29, the results indicate that, except for the case of the third-layer fifth-pass weld bead cross-section morphology where the binary segmentation method showed slightly better performance, the quadripartite segmentation method exhibited higher SSIM and larger PSNR in other scenarios. Overall, the quadripartite segmentation method demonstrates superior performance in handling welding bead section images compared to the binary segmentation method.

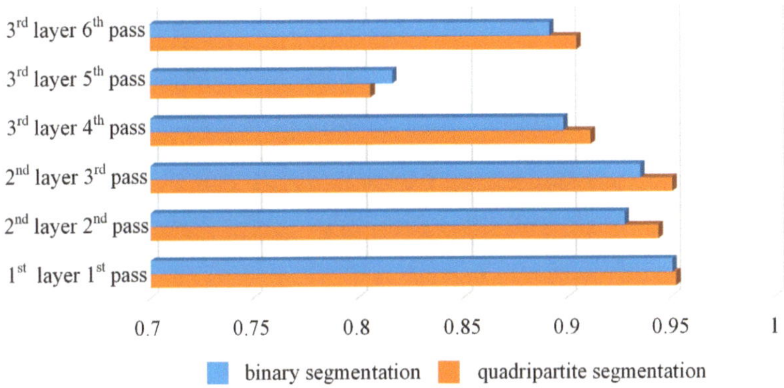

Figure 28. Comparison of SSIM.

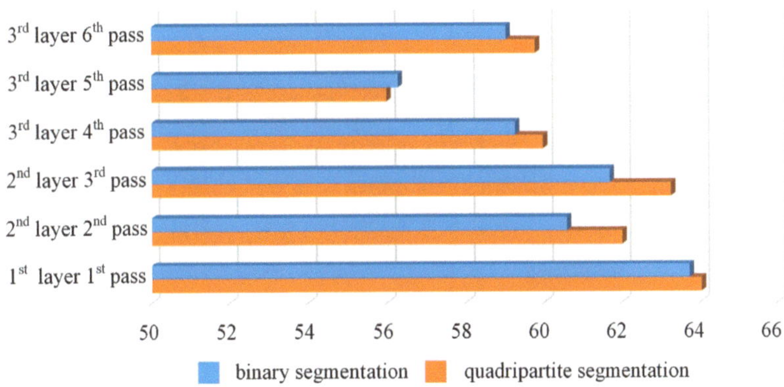

Figure 29. Comparison of PSNR.

This row–column-based grayscale image segmentation method provides an effective means to finely process weld bead images based on their local characteristics. Dividing the weld bead image into smaller regions can make more precise brightness adjustments for ESI, thus maximizing the preservation of weld bead details. This not only helps improve the overall visual quality of the weld bead image but also provides a more reliable foundation for subsequent image processing tasks.

5. Conclusions

(1) The innovation is the segmentation of the weld bead cross-section morphology into different modules based on the gray value for image processing. The ESI has an individual threshold, which allows the weld bead cross-section morphology to be extracted by region. The individually extracted morphology is then fused to obtain the complete weld bead cross-section morphology.
(2) A segmentation points extraction method based on the average gray value in the row and column is proposed. This method solves the uneven distribution of the gray value of the weld image after segmentation and can extract a more convenient environment for the subsequent processing.
(3) Multi-layer multi-pass weld bead cross-section morphology extraction experiments were conducted, ranging from first-layer first pass to third-layer sixth pass. The relative errors in WSC and WSA are within 10%, while the discrepancies in MWSW and MWSH can be close to the true value. The quality assessment falls within a reasonable range, which the average value of SSIM is superior to 0.9 on average and the average value of PSNR is superior to 60. This indicates that the multi-layer multi-pass weld feature extraction method based on image processing, while requiring a certain level of image quality, can identify the general contour of the weld bead and extract its shape parameters. We have reason to believe that in future work, more complex methods can be integrated to further enhance the accuracy and precision of the identification process, ultimately replacing manual identification.

Author Contributions: Conceptualization, methodology, validation and investigation by T.L. Writing original draft, data curation and formal analysis by S.G.; Supervision by C.W. All authors have read and agreed to the published version of the manuscript.

Funding: This research is supported by National Natural Science Foundation of China (52305416). Fundamental Research Funds for the Central Universities (2024IVA025). Hubei Province Postdoctoral Outstanding Talent Tracking Support Program (202401jc0012).

Institutional Review Board Statement: Not applicable.

Informed Consent Statement: Not applicable.

Data Availability Statement: The original contributions presented in the study are included in the article, further inquiries can be directed to the corresponding author.

Conflicts of Interest: The authors declare no conflicts of interest.

References

1. Sabatakakis, K.; Bourlesas, N.; Bikas, H.; Papacharalampopoulos, A.; Stavropoulos, P. Laser Welding of Dissimilar Cell Tabs: Extracting Physics Semantics from Infrared (IR) Emissions as Process Monitoring Data. *Procedia CIRP* **2024**, *121*, 222–227. [CrossRef]
2. Ai, Y.; Dong, G.; Yuan, P.; Liu, X.; Yan, Y. The Influence of Keyhole Dynamic Behaviors on the Asymmetry Characteristics of Weld during Dissimilar Materials Laser Keyhole Welding by Experimental and Numerical Simulation Methods. *Int. J. Therm. Sci.* **2023**, *190*, 108289. [CrossRef]
3. Dong, W.; Lian, J.; Yan, C.; Zhong, Y.; Karnati, S.; Guo, Q.; Chen, L.; Morgan, D. Deep-Learning-Based Segmentation of Keyhole in In-Situ X-ray Imaging of Laser Powder Bed Fusion. *Materials* **2024**, *17*, 510. [CrossRef] [PubMed]
4. Seong, W.J.; Park, S.C.; Lee, H.K. Analytical Model for Angular Distortion in Multilayerwelding under Constraints. *Appl. Sci.* **2020**, *10*, 1848. [CrossRef]
5. Cheng, W.; Zhang, X.; Lu, J.; Dai, F.; Luo, K. Effect of Laser Oscillating Welding on Microstructure and Mechanical Properties of 40Cr Steel/45 Steel Fillet Welded Joints. *Optik* **2021**, *231*, 166458. [CrossRef]
6. Fang, J.; Wang, K. Weld Pool Image Segmentation of Hump Formation Based on Fuzzy C-Means and Chan-Vese Model. *J. Mater. Eng. Perform.* **2019**, *28*, 4467–4476. [CrossRef]
7. Tohmyoh, H.; Ito, M.; Hasegawa, Y.; Matsui, Y. Extraction of the Outer Edge of Spot Welds from the Acoustic Image with the Aid of Image Processing. *Int. J. Adv. Manuf. Technol.* **2021**, *114*, 1731–1740. [CrossRef]
8. Hao, J.; Ding, M.; Li, Z.; Liu, X.; Yang, H.; Liu, H. Molten Pool Image Processing and Quality Monitoring of Laser Cladding Process Based on Coaxial Vision. *Optik* **2023**, *291*, 171360. [CrossRef]

9. Siva Shanmugam, N.; Buvanashekaran, G.; Sankaranarayanasamy, K. Some Studies on Weld Bead Geometries for Laser Spot Welding Process Using Finite Element Analysis. *Mater. Des.* **2012**, *34*, 412–426. [CrossRef]
10. Zhao, X.; Chen, J.; Zhang, W.; Chen, H. A Study on Weld Morphology and Periodic Characteristics Evolution of Circular Oscillating Laser Beam Welding of SUS301L-HT Stainless Steel. *Opt. Laser Technol.* **2023**, *159*, 109030. [CrossRef]
11. Wójcicka, A.; Walusiak, Ł.; Mroczka, K.; Jaworek-Korjakowska, J.K.; Oprzędkiewicz, K.; Wrobel, Z. The Object Segmentation from the Microstructure of a FSW Dissimilar Weld. *Materials* **2022**, *15*, 1129. [CrossRef] [PubMed]
12. Ai, Y.; Lei, C.; Yuan, P.; Cheng, J. Analysis of Weld Seam Characteristic Parameters Identification for Laser Welding of Dissimilar Materials Based on Image Segmentation. *J. Laser Appl.* **2022**, *34*, 042050. [CrossRef]
13. Zhang, B.; Shi, Y.; Cui, Y.; Wang, Z.; Chen, X. A High-Dynamic-Range Visual Sensing Method for Feature Extraction of Welding Pool Based on Adaptive Image Fusion. *Int. J. Adv. Manuf. Technol.* **2021**, *117*, 1675–1687. [CrossRef]
14. Lei, T.; Gu, S.; Yu, H. Keyhole Morphology Monitoring of Laser Welding Based on Image Processing and Principal Component Analysis. *Appl. Opt.* **2022**, *61*, 1492. [CrossRef] [PubMed]
15. Sun, J.; Zhang, C.; Wu, J.; Zhang, S.; Zhu, L. Prediction of Weld Profile of 316L Stainless Steel Based on Generalized Regression Neural Network. *Hanjie Xuebao/Trans. China Weld. Inst.* **2021**, *42*, 40–47. [CrossRef]
16. Alaknanda; Anand, R.S.; Kumar, P. Flaw Detection in Radiographic Weldment Images Using Morphological Watershed Segmentation Technique. *NDT E Int.* **2009**, *42*, 2–8. [CrossRef]
17. Jingbo, S.; Chengbin, S.; Jianye, Z. Study on Weld Penetration and Microstructure of Q345B Low Alloy Steel Plate by Electron Beam Welding. *Wide Heavy Plate* **2022**, *28*, 25–27.
18. Wu, X.; Su, H.; Sun, Y.; Chen, J.; Wu, C. Thermal-Mechanical Coupled Numerical Analysis of Laser + GMAW Hybrid Heat Source Welding Process. *Hanjie Xuebao/Trans. China Weld. Inst.* **2021**, *42*, 91–96. [CrossRef]
19. Cui, S.; Pang, S.; Pang, D.; Zhang, Z. Influence of Welding Speeds on the Morphology, Mechanical Properties, and Microstructure of 2205 DSS Welded Joint by K-TIG Welding. *Materials* **2021**, *14*, 3426. [CrossRef] [PubMed]
20. Peng, J.; Aimin, W.; Jin, Y.; Xinyi, S. Fast Recognition Algorithm of Weld Seam Image under Passive Vision. In Proceedings of the 2020 3rd World Conference on Mechanical Engineering and Intelligent Manufacturing, WCMEIM 2020, Shanghai, China, 4–6 December 2020; Institute of Electrical and Electronics Engineers Inc.: Piscataway, NJ, USA, 2020; pp. 351–355.

materials MDPI

Article

Microwave Hybrid Sintering and Soldering of Cu-Cr-W Composite Material for Reactive Power Breakers

Sorin Vasile Savu [1,*], Cristian Daniel Ghelsingher [2], Iulian Ștefan [1], Nicușor-Alin Sîrbu [3], Andrei-Angelo Midan [2], Ilie Dumitru [4], Ionel Dănuț Savu [1], Claudiu Nicolicescu [1] and Andrej David [5]

1 Department of Engineering and Management of Technological Systems, Faculty of Mechanics, University of Craiova, 200585 Craiova, Romania; iulian.stefan@edu.ucv.ro (I.Ș.); ionel.savu@edu.ucv.ro (I.D.S.); claudiu.nicolicescu@edu.ucv.ro (C.N.)
2 Doctoral School Academician Radu Voinea, Faculty of Mechanics, University of Craiova, 200585 Craiova, Romania; ghelsingher.cristian.j4s@student.ucv.ro (C.D.G.); midan.andrei.c3i@student.ucv.ro (A.-A.M.)
3 NRDI for Welding and Material Testing, 300222 Timisoara, Romania; asirbu@isim.ro
4 Department of Automotive, Transport and Industrial Engineering, Faculty of Mechanics, University of Craiova, 200585 Craiova, Romania; ilie.dumitru@edu.ucv.ro
5 Department of Water Transport, Faculty of Operation and Economics of Transport and Communications, University of Zilina, 01026 Zilina, Slovakia; andrej.david@fpedas.uniza.sk
* Correspondence: sorin.savu@edu.ucv.ro

Abstract: Over 60% of reported failures for reactive power compensation systems are given for damage to electrical circuit breaker contacts. This paper presents a study on the development of microwave technology for sintering of W–Cu–Cr alloys at 1012 °C for 65 min using 623.38 W microwave power, as well as microwave joining at 231 °C of the W–Cu–Cr composite material on body contact using 475 W microwave power for 55 s. The joined components were subjected to mechanical and electrical tests in accordance with ICE standards to validate the applied technology. Tests of connection–disconnection of the electrical contacts were carried out in accordance with the maximum number of disconnections allowed by the manufacturer (2 cycles/min): 25 s rest time and 5 s operating time under load. The components of the electrical contact after 111237 switches were analyzed under a microscope revealing a reduction of the damaged area by 27% compared with the original contact.

Keywords: soldering; microwave heating; thermal field; sintered material

1. Introduction

The switchers/breakers intended for connecting/disconnecting capacitive loads are designed and executed with an ambient operating temperature for encapsulated systems in the range of ±40 °C, and a minimum number of operations equal to 10^5 switches. However, under operating conditions, the active elements of the switching equipment experience overheating of the breaking contacts, thereby leading to their welding [1]. The main negative effect consists in unbalancing the capacitor bank, or not achieving the compensation stage. Moreover, battery imbalance results in an erroneous active/reactive energy measurement when contact welding occurs on only one phase. Failure to achieve the compensation step is directly reflected in the consumer's final energy costs. According to the annual report of SERTIM Electric [2], the failure to achieve reactive energy compensation is mainly due to the failure of the switching systems, as a result of the destruction of their contacts. Figure 1 presents the number of contact failures in 2023 reported in 800 installed units. According to the report, all systems had been installed at the consumers and had been functioning at normal temperature, humidity, and pressure. The number of failures is an absolute number, as recorded in 2023, and represents the number of breakers replaced after failure.

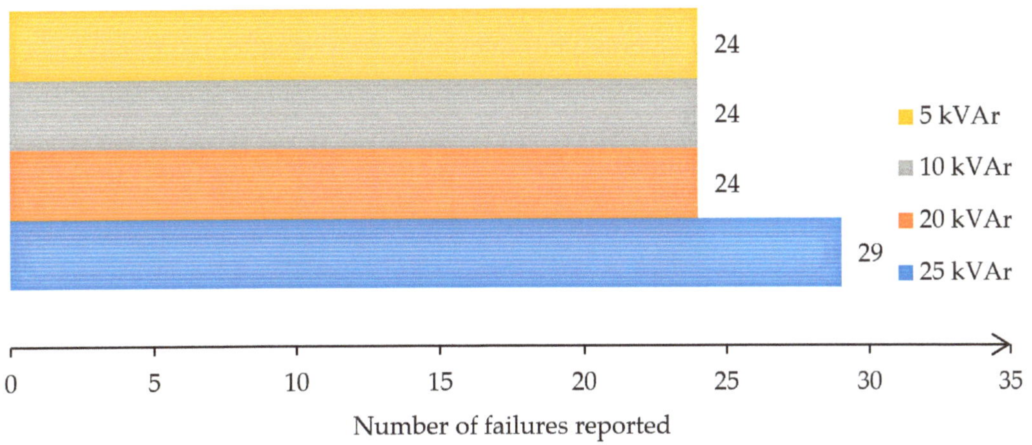

Figure 1. Failures of electrical contacts.

Microwave sintering is widely applied for ceramics, as well as for metals and composite materials. Dwivedi et al. [3] reported that the synergistic performance evaluation of copper–WC–graphene composites produced via microwave sintering has yielded promising results across various aspects. A microstructural analysis demonstrated a uniform distribution of reinforcing elements (WC and graphene) in the copper matrix for all compositions, highlighting the efficiency of the microwave sintering technique. Sun et al. [4] concluded that microwave hot press sintering led to a notable enhancement in the physical and mechanical properties of the composites, including their density and hardness. In addition, the graphite/copper composites exhibited both excellent thermal conductivity and frictional properties. Wang et al. [5] studied the thermal properties of graphite/copper composite material sintered in a microwave field. They reported that, with the increase in the number of thermal cycles, the thermal diffusion coefficient of the Gr/Cu composites decreased and gradually stabilized. By contrast, the CTE of the composites showed a linear increasing trend. Xu et al. [6] reported on the excellent performance of a W–Cu alloy sintered in a microwave field. The results obtained revealed a more uniform and compact microstructure with high density. Sustainable microwave processing has been reported by Kaushal et al. [7]. They concluded that processing materials/powders through microwave heating is a challenging task; yet, it is promising owing to its exceptional rapid heating characteristics, coupled with its potential energy savings and sustainability nature. The composite castings of Cu–W_x-based metallic powders were successfully processed through MHH at 2.45 GHz and 900 W. Cursaru et al. [8] studied the sintering process of ceramics for biomechanical applications, referring to microwave irradiation as a sustainable technology for materials sintering. The microwave sintering of materials has limitations, mainly relating to process stability. Stefan et al. [9] studied the thermal runaway phenomenon during the microwave heating of composite materials. They reported that ceramic materials tend to suffer from thermal runaway after 600 °C; this was the case even with low levels of microwave-injected power. Savu I.D. et al. [10] and Savu S.V. et al. [11] concluded that thermal runaway can be avoided by applying a simultaneous reduction of microwave-injected power down to 630 W and a manual control of the matching load impedance to increase the reflection of microwaves but ensuring, at the same time, the protection of the generator antenna. Thermal runaway has been studied widely [12–14], and is considered the most influential factor that creates instability during microwave heating.

In addition to microwave sintering, fabrication processes, such as welding, bonding and soldering, can be applied in a microwave field. Zhang et al. [15] investigated the heat transfer from ceramic susceptor to copper materials with SAC305 solder in microwave hybrid soldering. They concluded that cylindrical susceptors provide a more efficient and consistent heating effect compared to cubic susceptors. In addition, when the susceptor's height is 30 mm, the heating efficiency peaks, and the hot spot is located at the bottom of the susceptor. Said et al. [16] performed a review of microwave hybrid heating (MHH) providing understanding of MHH for an alternative soldering approach by compiling all the most recent research, technical setup, and key discoveries. The MHH soldering method shows remarkable performance for improving the strength (44.4%) of the solder joint compared to the conventional method. Savu S.V. et al. [17] studied the microwave soldering of copper plates for 600, 650, 700, and 750 W at 2450 MHz for a soldering time up to 60 s. They concluded that microwave soldering does not significantly influence the total electrical resistance which is mandatory for electronic applications. Other researchers have studied the grain orientation and failure mechanism of isothermal aged Cu–Cu joints using microwave hybrid heating [18]. Zhang et al. [19] reported that the interface of the MHH joints was larger than that of the reflow soldering joints obtained under the parameters of 195 s at a power of 900 W for process parameters at a peak temperature of 250 °C and a reflow time of 600 s.

This research aims to apply microwave technology for joining W–Cu–Cr composite material and contact body material using hybrid microwave heating. This study is also focused on microwave joining process stability, as well as testing the switching devices according to ICE standards to validate the joining technology and performance of the final product.

2. Materials and Methods

To develop the composite material, a mass composition (percentage by weight) of 94% Cu, 1% Cr, and 5% W was used. The powders used for the development of the composite material were as follows: copper powders, produced by Pometon S.p.A, Maerne, Italy; tungsten powders, produced by Electroputere Craiova SA, Romania; and chromium powders, produced by Merck KGaA Sigma Aldrich, Germany. The powders were weighed using an analytical balance: Partner WPS 510/C2, produced by Partner Corporation (London, United Kindom, 2008). The characteristics of the elemental powders are presented in Table 1.

Table 1. Main characteristics of elemental powders.

Parameter/Powder	Cu	Cr	W
Apparent Density	2.30–2.50 g/cm^3	7.14 g/cm^3	18.5 g/cm^3
Purity	99.7%	99.8%	99.1%
Mesh size	180–212 μm	70 μm	80 μm
Grain shape	dendritic	irregular	polyendric

These powders were mechanically alloyed using a planetary ball mill: Pulverisette 4, manufactured by Fritch (Idar-Oberstein, Germany, 2010). The technological parameters of the mechanical alloying process are presented in Table 2.

Table 2. Technological parameters of mechanical alloying process of W–Cu–Cr powders.

Parameter	Value
Bowl volume	250 mL
Ball diameter	10 mm
Ball material	stainless steel
Alloying environment	dry inert gas (Ar)
Ratio material/balls	1/2
Rotation speed	5.83 1/s on primary drive 8.33 1/s on bowl
Alloying time	5 h

2.1. Modelling and Simulation of Temperature Distribution at Microwave Processing

Microwave sintering of metal-based composite materials cannot be performed due to the high reflection of microwaves by metals [20]. Therefore, indirect microwave heating or hybrid microwave heating can be applied, but only after a simulation of temperature distribution to establish the lowest level of microwave power. Due to the reflection coefficient of metal, high levels of microwave power will lead to a plasma arc discharge and damage of the magnetron antenna. Therefore, to achieve a better conversion rate of the microwaves into heat, ceramic susceptors must be used. This approach defines the indirect or hybrid microwave heating, and usually is taken into consideration for the sintering of materials with a high reflection of microwaves. The heat developed by ceramic susceptors will be transferred based on Fourier's heat conduction law, as presented in Equation (1):

$$Q = -k \cdot A \cdot \frac{\Delta T}{h} = -k \cdot A \cdot \frac{(T_h - T_c)}{(h_e + h_i)} \tag{1}$$

where A (mm^2) is the heat transfer area, T_h (°C) is the highest temperature obtained from the conversion of the microwaves into heat, T_c = 25 °C is the ambient temperature considered for simulation, and $h = h_e + h_i$ = 3.22 mm represents the total length of the assembly and k (W/mK) represents thermal conductivity. Figure 2 presents the temperature distribution during microwave hybrid heating.

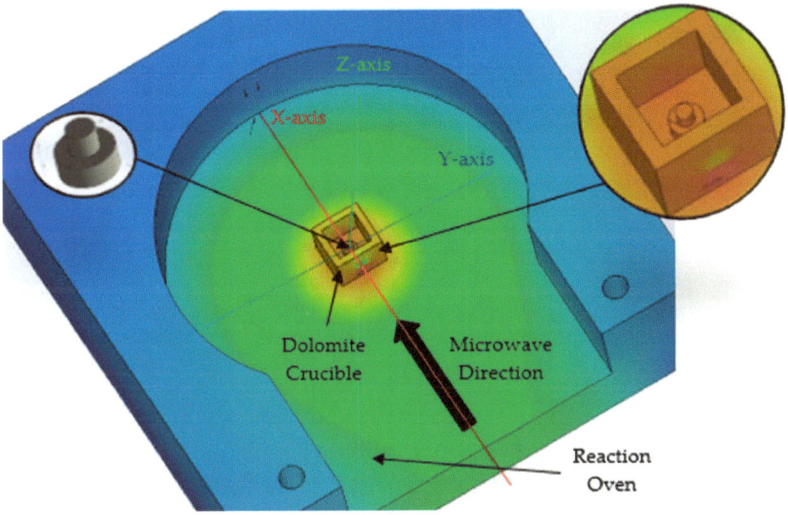

Figure 2. The model developed for temperature distribution on microwave indirect heating.

The model contains the microwave reaction chamber and a ceramic crucible (dolomite material) with overall dimensions 14 × 14 × 10 mm having a 1 mm-thick wall. The composite material, CuCr1W5, has a specific shape for active elements of the reactive switching devices, consisting of two cylinders with different diameters and heights, as follows: d_e = 4 mm, h_e = 1.60 mm; and d_i = 2 mm, h_i = 1.62 mm. Considering the simulation after the *X*-axis and *Y*-axis, the distribution of the temperatures are presented in Figure 3. The simulation has considered the following parameters and boundary conditions related to the heat transfer into solids: the heat conduction on the *Z*-axis is higher towards the bottom of the reaction oven, taking into consideration the contact between the dolomite crucible and the aluminum body of the oven. The heat transfer from the dolomite (indirect microwave heating) to the composite material CuCr1W5 and the reaction oven body uses specific categories of thermal parameters, such as thermal conductivity (k), heat expansion coefficient (α), and specific heat (c), is presented in Table 3 [21,22].

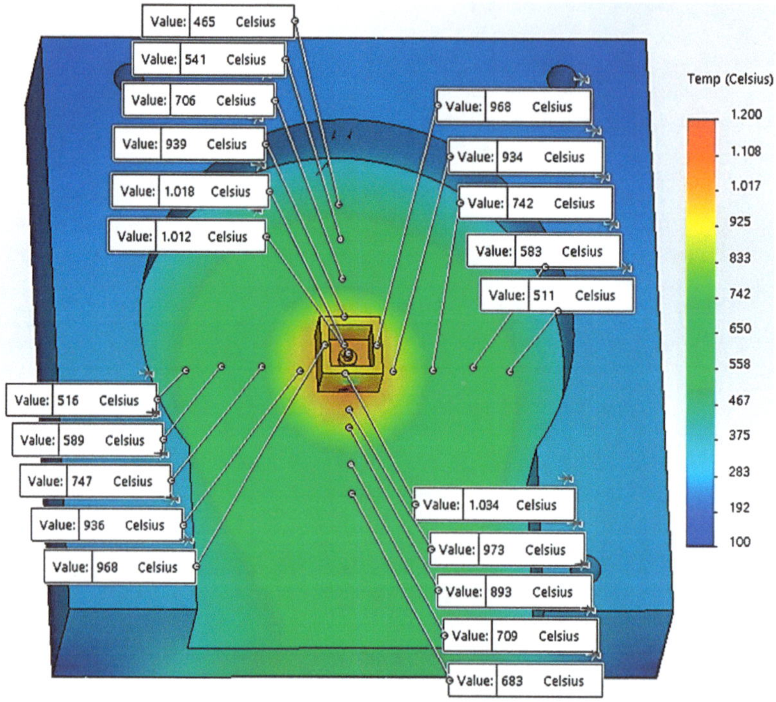

Figure 3. Temperature distribution on Y-axis and X-axis.

Table 3. Heat properties and coefficients of materials.

Material	k (W/mk)	α (10^{-6} m/m°C)	c (kJ/kgK)
Aluminum	231	22	1042
Dolomite	5.3	4.6	0.92
CuCr1W5	398	17	390

According to the simulation of thermal transient for 5 min, in Figure 3, the temperature distribution, at the contact between the dolomite crucible and the oven body was approximately 930 °C for both the *X*- and *Y*-axis, suggesting a uniform distribution inside the crucible. Similarly, on the *Z*-axis, the simulation of temperature reveals that at the top of the sample the value obtained was 1012 °C.

The microwave soldering of copper–copper plates has been studied in previous research [17]. The results have shown that various levels of microwave-injected power from 650 W to 750 W can be applied to achieve a stable soldering process in a microwave field with good quality of joints, and with a low, indirect carbon footprint and costs. However, the microwave soldering of composite sintered material based on CuCr1W5 with a contact body requires study on the microwave parameters to achieve a uniform distribution of temperature in the soldering area. Therefore, before soldering in a microwave field, a simulation of temperature distribution is required. That must be studied due to the high reflection coefficient of microwaves of metals. Similarly to microwave sintering, for a joining application, it is necessary to use ceramic susceptors in order to convert microwaves into heat. The model for the simulation of temperature distribution, presented in Figure 4, was performed for a contact body having 15.70 × 10.50 mm dimensions; an active element having d_e = 4 mm, h_e = 1.60 mm and d_i = 2 mm, h_i = 1.62 mm; a solder SAC305, with chemical composition SnAg3Cu0.5 produced by Cynel Unipress (Warsaw, Poland, 2023), deposited on contact surfaces; and a dolomite susceptor, with a parallelepiped shape having the overall dimensions 50 × 50 × 40 mm.

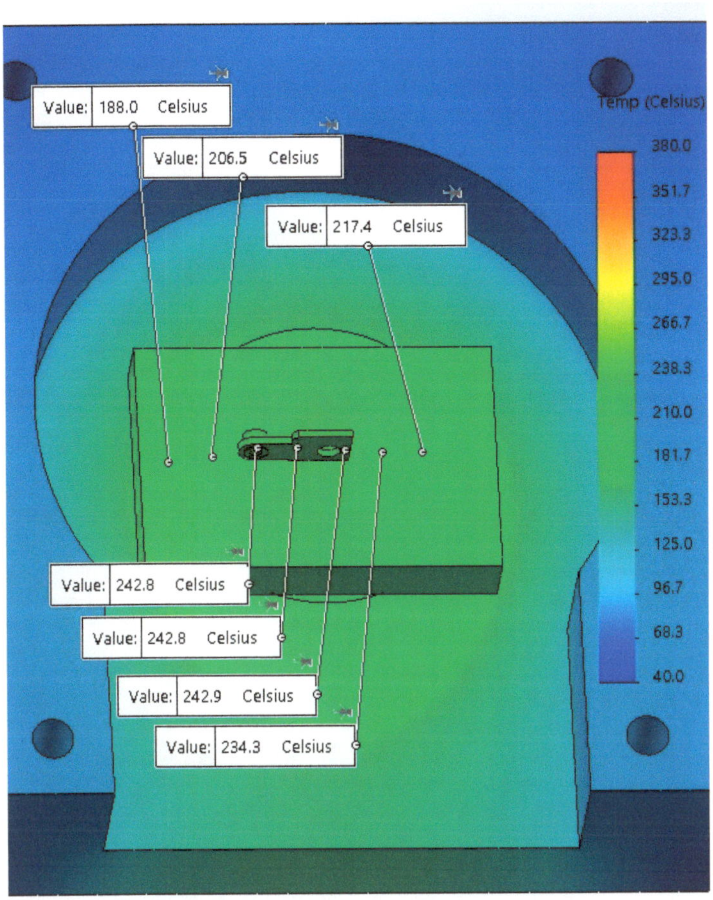

Figure 4. Temperature distribution during microwave soldering process.

The active element of the electrical contact was designed and executed exactly like an industrial contact existing in a reactive switch, type CEM 5 kVAr, produced by ETI Elektroelement d.o.o. (Izlake, Sloven Republic, 2024).

2.2. Microwave Sintering of Composite Material

Based on the simulation results, the temperature developed inside the ceramic susceptor (dolomite crucible) was evaluated to be approximately 1013 °C; this will be used to determine the necessary microwave-injected power using the following Equation (2):

$$P_{MW} = k \cdot V \cdot f^2 \cdot \varepsilon \cdot \tan\delta \cdot \frac{\Delta T}{t} \quad (2)$$

where P_{MW} (W) is the microwave-injected power from the generator, $k = 1.01 \times 10^{-9}$ is the proportionality constant depending on the geometry and material properties of the crucible [23], $V = 0.0000098$ m^3 is the overall volume of dolomite crucible, f = 2450 MHz is the frequency of microwave radiation, ε = 7.2 F/m [24] is the permittivity of dolomites,

$\tan\delta = 4.32 \times 10^{-4}$ at 2450 MHz [25] is loss tangent of dolomites, $\Delta T = 1012$ °C is the temperature required at the contact between crucible and sample material, and $t = 300$ s is the total time required for reaching the sintering temperature of the composite material. By solving Equation (2), the minimum level of injected power has been determined to be $P_{MW} = 623.38$ W.

By using the calculated microwave power, the experimental program for sintering the composite material based on CuCr1W5 was performed using a microwave installation, as presented in Figure 5.

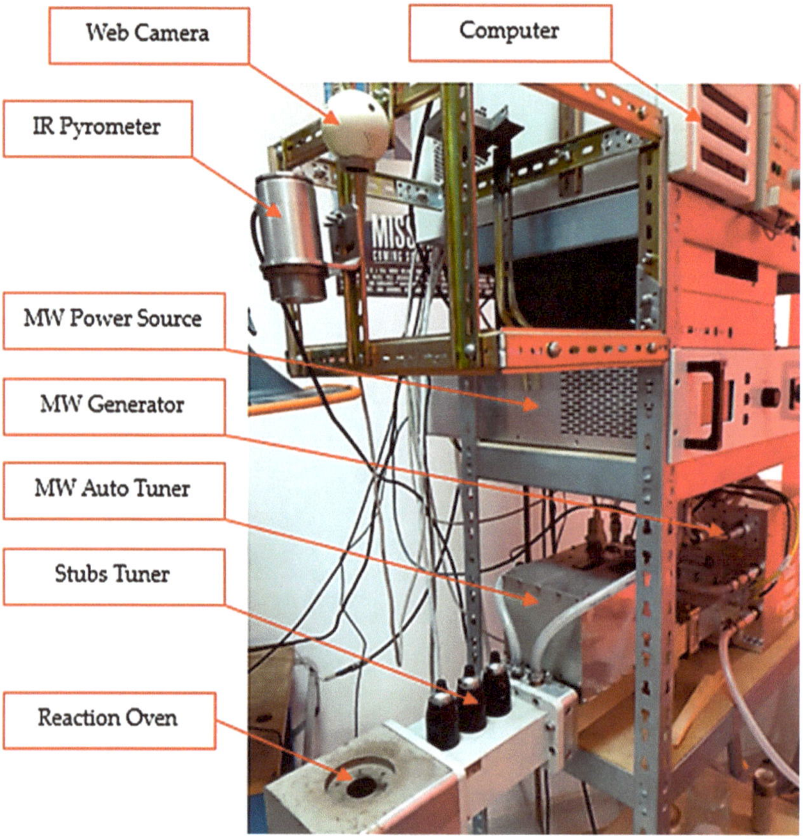

Figure 5. Experimental microwave soldering installation.

The microwave sintering procedure of the composite material was supported by a microwave heating installation consisting of an MW-Generator Set 6000 W, 2450 MHz continuous wave containing a magnetron head type MH6000S-251BF, from Muegge GmbH (Reichelsheim, Germany, 2020). The microwave generator was driven by an MW-Power Source Supply 6000 W, 2450 MHz, 3 × 400 V, continuous wave type MX6000D-154KL, both manufactured by Muegge GmbH (Germany, 2020). The main auxiliary devices were an infrared pyrometer CSmicro 3M manufactured by Optris GmbH (Berlin, Germany, 2020), having a temperature range between 50 °C to 600 °C, with a spectral range of 2.3 μm, used for temperatures below 600 °C; and a CT Glass G5H manufactured by Optris GmbH (Germany, 2012), having a temperature range between 250 °C to 1650 °C, with a spectral range of 2.1 μm, used for temperatures above 600 °C. Both devices are ideal for the temperature measurement of metallic surfaces. The control of the infrared pyrometer was performed using a Compact Connect supplied by the manufacturer together with an IR pyrometer. In addition, for matching the load impedance, a Tristan auto tuner up to 6000 W, 2450 MHz, from Muegge GmbH (Reichelsheim, Germany, 2008), was used. The tunning process, related to matching the load impedance, was supported by Homer Software, provided by S-Team d.o.o. (Bratislava, Slovakia, 2008).

The microwave process parameters related to matching the load impedance to the microwave generator, in addition to the injected power, was determined experimentally through the microwave heating of the dolomite crucible without any loads. The values of the stub tuners were calculated and then recorded using the Tristan auto tuner and Homer Software. The process was set for high temperatures up to 1400 °C to be able to analyze the behavior of dolomite in a microwave field without injected powers higher than 600 W. The microwave sintering was performed in a normal atmosphere, but a jet of inert gas (argon) was directed to the samples with a flow rate of 2 L/min. Figure 6 presents the temperature evolution in correlation with the injected power for the resonant circuit.

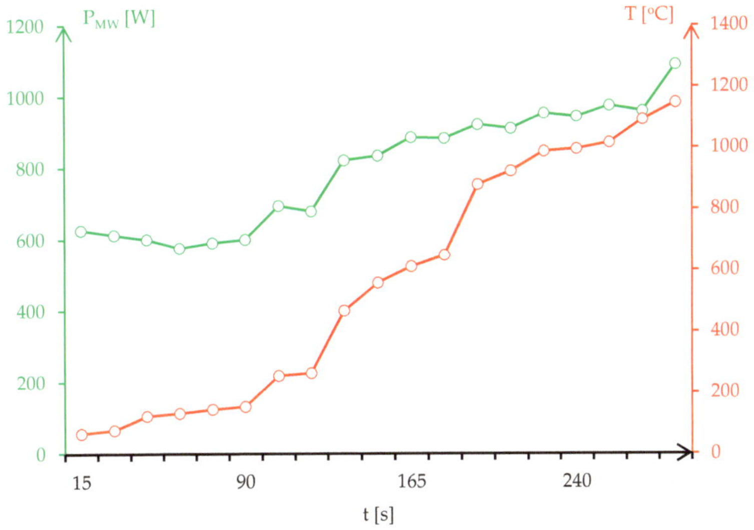

Figure 6. Temperature evolution on the surface of the composite material as a function of the injected power.

Once the presumed sintering temperature T = 1012 °C had been reached, the heating process was maintained for 60 min. The sintered sample was then prepared for microscopy analysis.

2.3. *Hybrid Soldering of Sintered Active Element to the Contact Body*

Based on a simulation of the temperature distribution, the necessary microwave-injected power for soldering was calculated using Equation (2) to 533.25 W, taking into consideration the solder SAC305 having the thickness 0.1 mm. This level of power was reached in an experimental application after the calculation of the distance of stub tuners using Equation (3):

$$d_i = -\frac{\lambda}{2\pi}\cdot\tan^{-1}(B_i\cdot Z_0) \quad (3)$$

where d_1, d_2, and d_3 (mm) represent the length of the three stub tuners inside the WR340 waveguide, $\lambda = 0.122$ m is the wavelength of microwaves at 2450 MHz, the susceptance of stub tuners being considered $B_1 = B_2 = B_3 = 0.5$ mS, and $Z_0 = 376.7\ \Omega$ is the air impedance. The computation of Equation (3) for the assembly of the susceptor-contact-solder-sample led to the following values: $d_1 = 0$ mm, $d_2 = 22.62$ mm, and $d_3 = 24.84$ mm. The results for stub tuners was validated by the stubbing scenario introduced in Homer Software for manual tunning of microwave heating process. Figure 7 presents the stub tuner movements for resonant circuit and the absorbed and reflected power during the microwave soldering process.

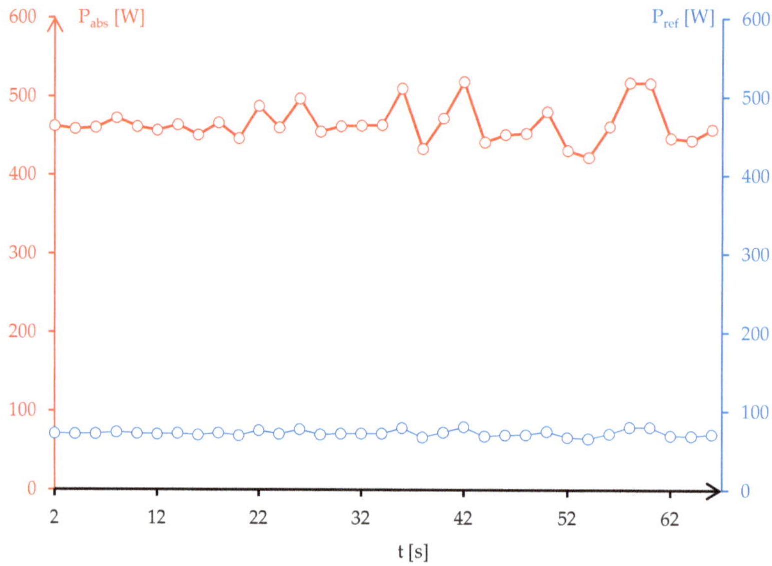

Figure 7. Resonant circuit for microwave soldering of sample to the contact body.

The experimental process of microwave soldering was started using 533 W representing 8.83% of the total available power that can be supplied by a microwave generator. At this level of power, the process tends to be unstable and difficult to control. However, the stability of dolomites in the microwave field led to a rapid increase in temperature. After 55 s, the temperature recorded on the surface of the sample was 231 °C, enough to perform the joining of the sample to the contact body. This level of temperature was obtained at 475 W, according to a three-stub tuner position and the microwave power absorbed, which was lower with 12% of the calculated microwave power using Equation (3). The difference can be explained due to the presence of SAC305 that became liquid after 217 °C, which improved the heating mechanism conditions and led to the better conversion of microwaves into heat.

2.4. *Testing New Electrical Contact*

The new electrical contact developed through the microwave sintering of composite material based on CuCr1W5 and soldered in a microwave field to the contact body was tested according to ICE standards for more than 100,000 cycles of breaking reactive circuit. In order to monitor the performance of the breaking device, a system composed of a three-phase reactive capacitor, a digital multimeter for reactive current measurement, a GSM module for the transmission of data measured and the status of the breaker, as well as an integrated circuit for time record has been developed. Figure 8 presents the schematic diagram of the experimental device.

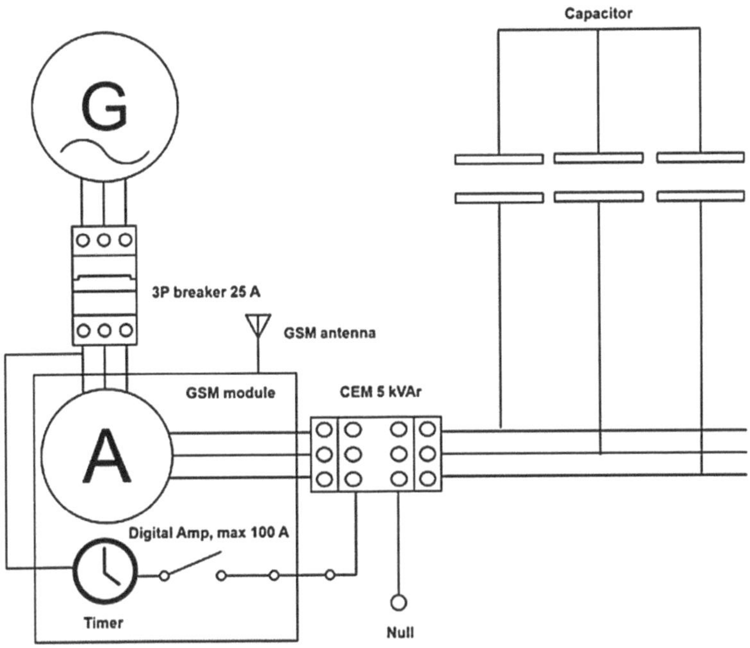

Figure 8. Schematic diagram of experimental device designed for breaking test.

An experimental device has been designed to test the functionality and performance of the new contact for breaking reactive loads. The system consists of a main breaker type 3P PL6–25/3, produced by Eaton (Kearney, NE, USA, 2024), connected to an electrical grid having a nominal breaking current up to 25 A. The main part of the system is composed of a digital multimeter for the measurement of AC currents up to 100 A, using Hall sensors ACS758LCB-050U-PFF-T produced by Allegro MicroSystems (Musselburgh, United Kingdom, 2024), a timer for breaking/connecting cycles integrated, and a GSM module connected to an external antenna both being built in a development motherboard type MKR 1400 GSM manufactured by Arduino LLC (Firenze, Italy, 2024). The functionality of the system is driven by a controller programmed to transmit information in real time in terms of AC current values for each phase. Thus, the breaking circuit CEM 5 kVAr produced by ETI Elektroelement d.o.o. (Izlake, Slovenia, 2024) containing the new contact based on CuCr1W5 will be monitored individually during the breaking cycle. The active part of the system is composed of a breaking circuit and a three-phase capacitor type LPC 3 × 3.33 μF, 5 kVAr, 400 V, 7.2 A connected as a load. Figure 9a presents the manufactured device produced in the scope of the research and Figure 9b presents the breaking/connecting cycle established to evaluate the performance of the new contact.

The testing of functionality and the performance of new electrical contact using composite material based on CuCr1W5 was performed for 111,237 connecting/breaking cycles until the contacts were welded. The information related to the status of the relay contacts was submitted automatically by the MKR 1400 GSM board, manufactured by Arduino LLC (Firenze, Italy, 2024). The monitoring process was not focused only on the failure of the switch/breaker device, but also on the values of the capacitive currents.

The new contact was embedded in the middle (S-phase of 3P system) of the three-phase relay contacts, the other two contacts being originals from the manufacturer. Before embedding the contacts in breakers, the samples were cleaned with alcohol to decontaminate the surface of the active elements. In addition, before installing the contacts into the breaker, the electrical resistance of the samples-contact body was measured using a 4-wire method. The results obtained were as follows: $R_R = 0.065\ \Omega$ for R-phase, $R_S = 0.055\ \Omega$ for S-phase, and $R_T = 0.060\ \Omega$ for T-phase. The capacitive currents measured before failure of the system were as follows: $I_R = 2.27$ A for R-phase, $I_S = 2.55$ A for S-phase, and $I_T = 2.42$ A for T-phase. From the beginning, the system was unbalanced by 11.07% due to the electrical connections and the performance of the capacitor. After 111,237 cycles, the system reported a failure on T-phase due to the welded contacts.

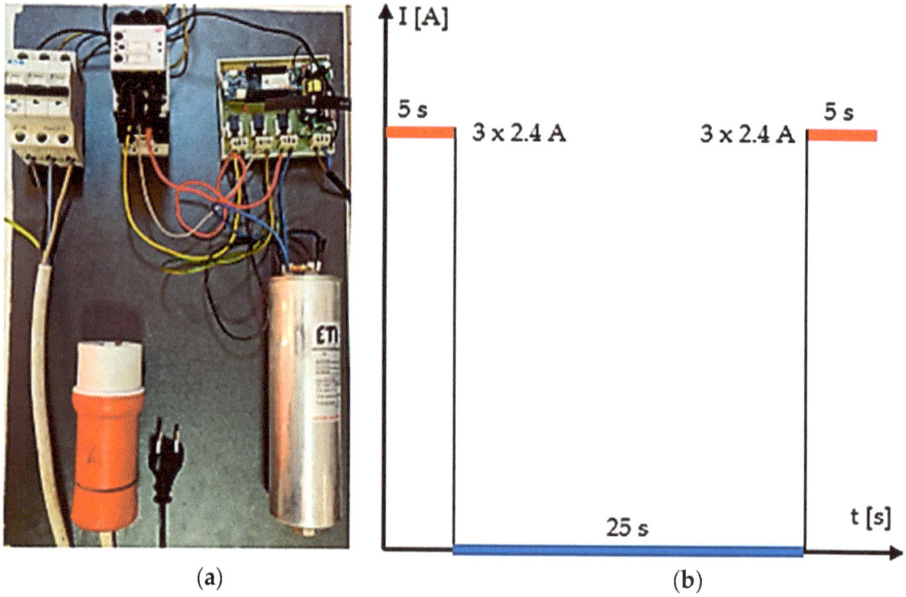

Figure 9. Experimental program for testing the new contact: (**a**) experimental device, (**b**) breaking cycle.

In addition, to evaluate the microhardness of the sintered composite material, a Namicon CV-400 DTS tester produced by CV Instruments (Thornaby, United Kingdom, 2007) was used. The microhardness tests were performed in accordance with EN ISO 6597 standards using Vickers method with the following process parameters: peak angle of indenter—136°, pressing force 4.9 N (HV0.5), and dwell time 15 s. The microhardness results obtained after 3 tests are presented in Table 4.

Table 4. Microhardness of composite material.

Test Parameters	Test 1	Test 2	Test 3	Average
Vickers method Tip angle 136° Load 4.9 N (HV0.5) Dwell time 15 s	45.4	51.9	48.5	48.6

After the testing process of the contacts, the damaged surfaces of the samples were measured using a Surtronic 25 profilometer, Taylor Hobson (Leicester, United Kingdom, 2007), with the following characteristics: calibration limit—300 µm, resolution—0.01 µm, examination length—0.25–25 mm, examination speed—1 mm/s, and computer connection RS232. The profile images are presented in Figure 10.

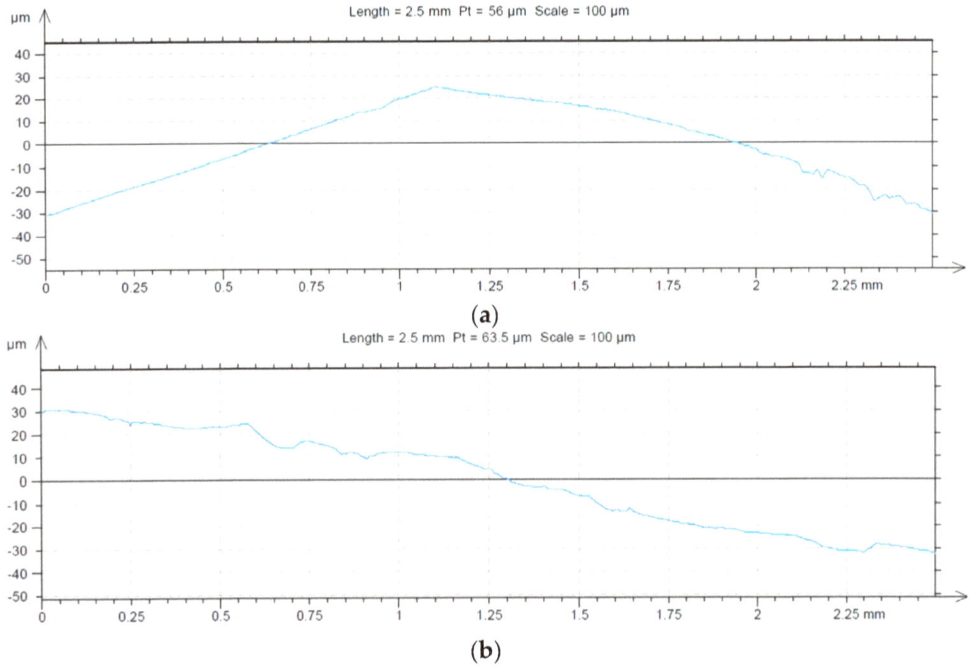

Figure 10. Profile images of damaged surfaces: (**a**) original contact based on copper coated with silver, (**b**) composite material based on CuCr1W5.

3. Results and Discussions

The development of a new contact based on an active element from the composite material CuCr1W5 has been performed, starting from reports [1,2], related to the level of malfunction of the electrical contacts embedded in reactive/capacitive load breakers. The development of a new contact based on the composite material CuCr1W5 has been performed using an MHH process for sintering. The results related to the microstructure of the sample are presented in Figure 11. According to Figure 11b, the grain size of tungsten is 25 µm, the chromium grain size is below 5 µm, and the grain size of the copper matrix is also below 25 µm. The structure of composite material seems to be uniformly distributed.

Regarding the soldering of the sintered composite material and the copper contact body, and the low interval of the soldering temperature (217–228 °C for SAC305 solder) is not affected by the thermal runaway phenomenon, but the instability of the microwave soldering process must be addressed. The instability of the process can be observed in the balance of power during microwave heating. After 20 s of heating, the reflected microwave power P_{ref} [W] oscillates between 66 W and 79 W, which represents more than 10% of the total injected power. For small applications, this level of reflected power does not represent an issue; but, for multiple soldering processes (wide scale productions similar to industrial processing) in the same reaction oven, the loss of microwave power will lead to a low quality of the joints.

In addition to Figure 7 and according to Figure 12, the matching load impedance of the assembly metal sample-susceptor is difficult to achieve due to the presence of the metal in the microwave field. However, the circuit impedance remains at low values mainly due to the scenario with the position of the first stub tuner to null (d_1 = 0 mm), which allows the passing of microwaves from the generator to the load. In addition, the calculated positions for the other stub tuners (d_2 = 22.62 mm, d_3 = 24.84 mm) do not allow the reflected microwave to pass the waveguide WR340 back to the microwave generator.

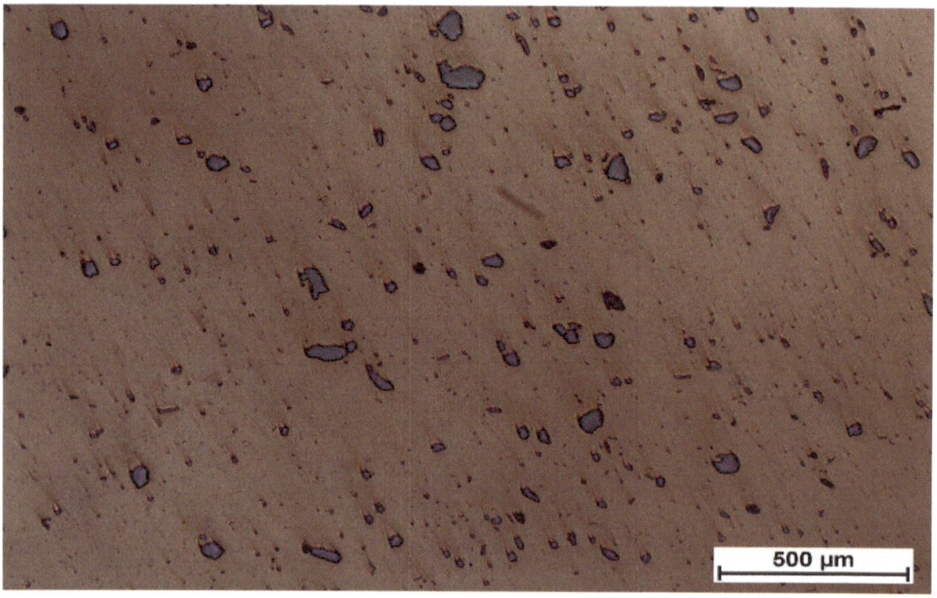

(**a**)

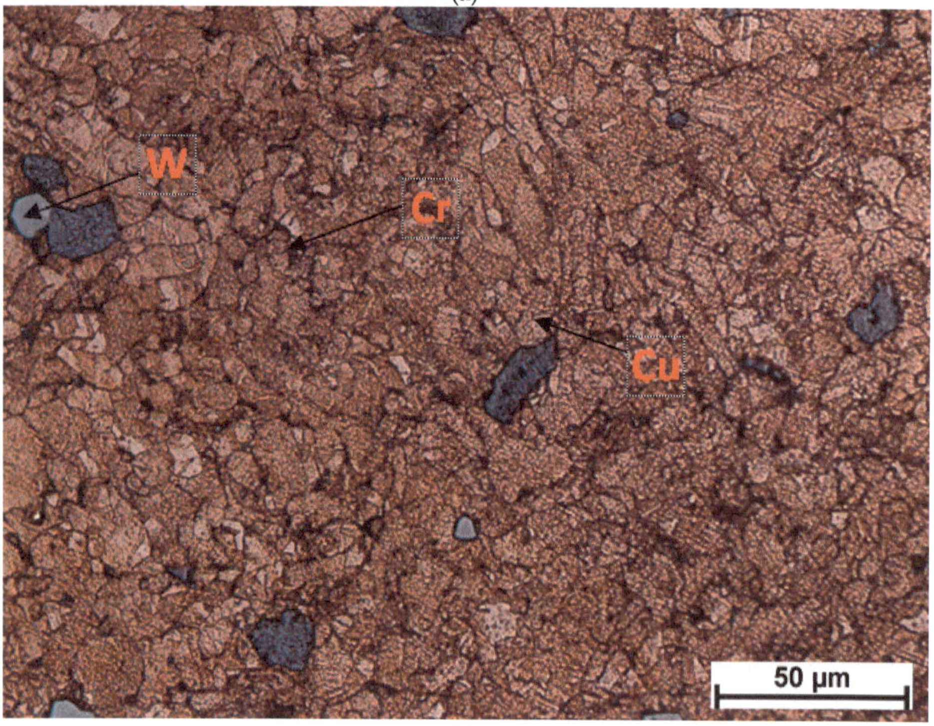

(**b**)

Figure 11. Microstructure of sample: (**a**) non etched (just polished) 75×, (**b**) etched (cupric chloride) 750×.

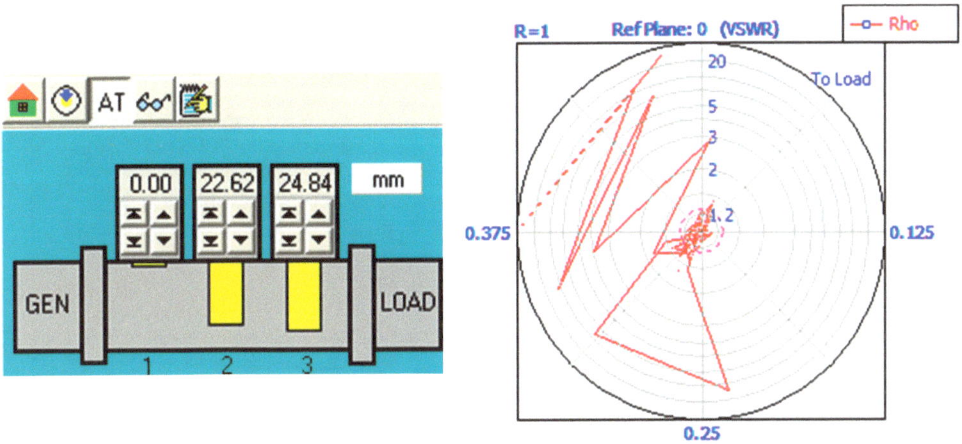

(a) (b)

Figure 12. Matching load impedance of assembly CuCr1W5—dolomite to the microwave generator during the soldering process: (**a**) stub tuners position, (**b**) absorbed/reflected microwave (courtesy Homer Software V5.2).

The testing of new contact functionality and performance has been realized using an experimental device. After 111,237 cycles, the system reported an error T-phase on the breaking cycle. This led to the conclusion that the contact suffered during the temporary welding process and that breaking was no longer available, for a short time, for all three phases.

After suspending the testing process, the electrical contacts have been analyzed under a microscope to evaluate the joint quality and to identify the defects. Figure 13 presents a comparative analysis of the damaged area for the T-phase and S-phase contacts.

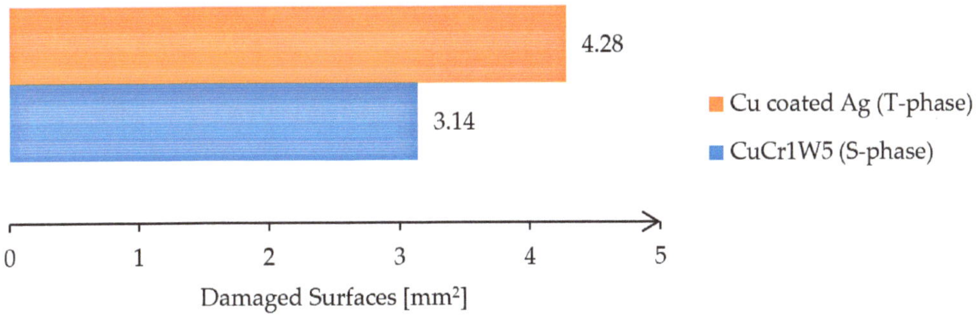

Figure 13. Areas of damaged surfaces of the contacts.

The chart presented in Figure 13 shows that the damaged area of the composite material based on CuCr1W5 is around 3.14 mm^2, being 27% smaller than the damaged area, around 4.28 mm^2, recorded for the original contact based on copper coated with silver. This can be explained by the properties of tungsten (a high melting point around 3422 °C and a high thermal conductivity) related to its level of degradation at a high temperature as well as its behavior under high AC currents. This behavior under high AC currents consists, of allowing it to withstand extreme heat without melting or deforming, is due to tungsten's high melting point. In addition, tungsten has a relatively low electrical resistance, meaning it can carry large currents efficiently but without significant erosion or degradation. On the other hand, according to Figure 14a, the new contact was affected mainly at the edge of the surface, which can be explained by the lack of contact on the entire surface. An analysis of the structure on the surface of the contacts used in the experimental program is presented in Figure 14.

The macroscopic examination of Figure 14b reveals that the electrical contact surface experienced electro-erosion, likely due to electric arcs forming during the closing or opening of the contacts. The effect is double: a thermal stress producing oxidation on about 20% of the surface, including oxides of copper, tungsten, and chromium, indicating exposure to an open-air environment, and a material removal, with excavation depths ranging from 21 to 56 μm, possibly caused by imperfections like microcracks. Thermal stress and its effect of thermal expansion produced oxidation on about 20% of the surface, including oxides of copper, tungsten, and chromium, indicating exposure to an open-air environment. The presence of the oxide types has been determined using an EDS analysis performed using TESCAN VEGA LMU equipment produced by TESCAN Brno s.r.o (Brno, Czech Republic, 2023). The results are presented in Figure 15. Regarding the original contact, the access of oxygen to copper is in the areas in which the silver is removed by extremely fast vaporization under electric discharge. But the entire silver layer is not removed by vaporization, and oxides of silver are present even in the affected area. The authors know that the potential of copper to oxidize is higher, but silver has the potential at the highest exposure to oxygen.

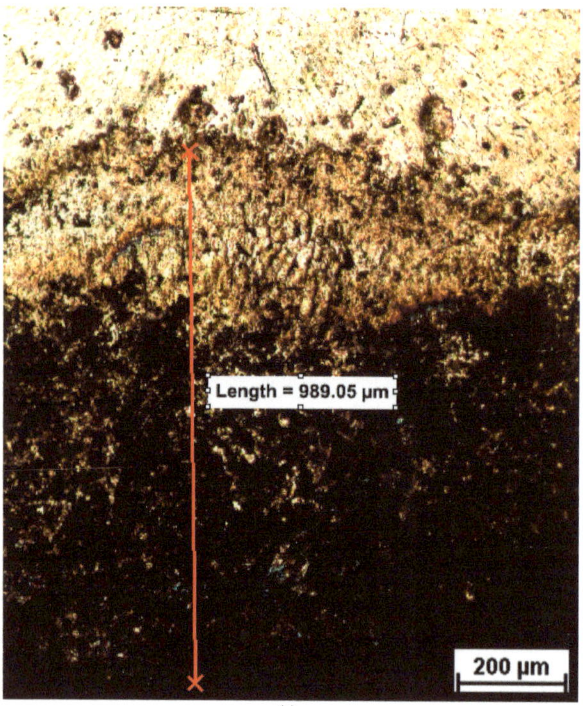

(a)

Figure 14. *Cont.*

(**b**)

Figure 14. Surface structure analysis: (**a**) S-phase contact for composite material based on CuCr1W5, (**b**) T-phase contact for original contact based on copper coated with silver.

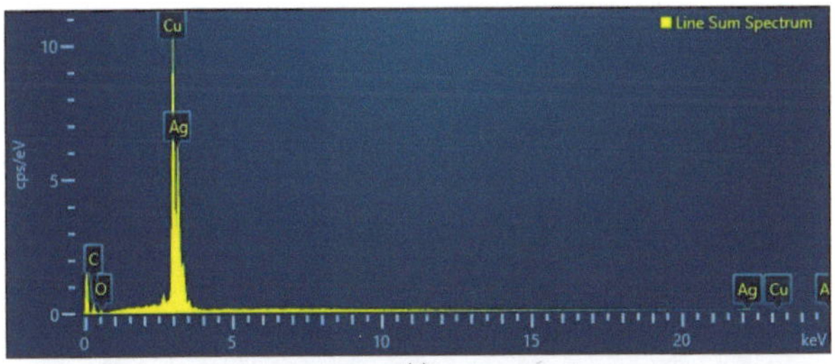

(**a**)

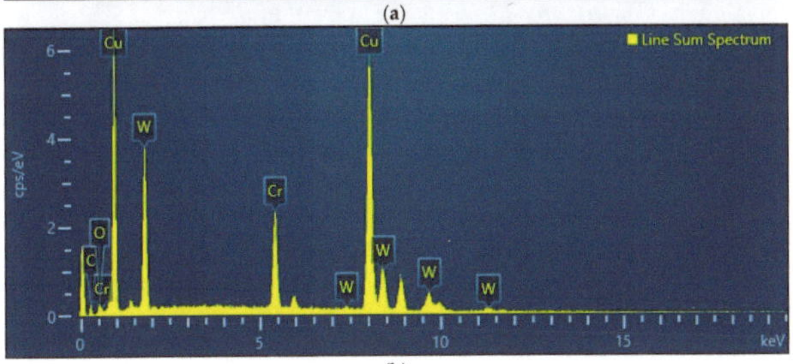

(**b**)

Figure 15. EDS analysis: (**a**) original contact based on copper coated with silver, (**b**) new contact from composite material based on CuCr1W5.

Regarding the composite material, the order of potential of oxidation is: copper, followed by chromium and tungsten. Oxygen is present and it has access to copper and chromium under high temperatures, producing a large quantity of copper oxides and chromium oxides in the affected area by discharge.

4. Conclusions

The lifespan of electrical contacts can be extended by using composite materials in relay contacts. The composite was sintered using hybrid microwave heating with dolomite-based susceptors, as direct microwave heating is challenging due to the low absorbance of metal. Simulations determined that the optimal microwave settings to achieve a sintering temperature was 1012 °C with 623.38 W of power. The sintering process took 65 min in total, which is much faster than conventional methods.

The composite material was then soldered to a copper contact body, with simulations again guiding the process. A soldering temperature of 231 °C was reached in 55 s using 475 W of power, making it more efficient than traditional soldering.

The new electrical contact was tested according to ICE standards, using a device designed to handle a 5 kVAr capacitive load. After 111,237 cycles, an error was detected in the T-phase due to contact welding.

Author Contributions: Conceptualization, S.V.S. and C.D.G.; methodology, I.D.S. and I.D.; software, I.Ș. and A.-A.M.; validation, I.D.S., I.Ș., C.N., and C.D.G.; formal analysis, I.D., A.D., and N.-A.S.; investigation, S.V.S., C.D.G., A.-A.M., and C.N.; resources, I.D., A.D., and N.-A.S.; data curation, I.D.S.; writing—original draft preparation, S.V.S.; writing—review and editing, I.D.S.; visualization, N.-A.S. and A.D.; supervision, S.V.S. All authors have read and agreed to the published version of the manuscript.

Funding: This research received no external funding and the APC was funded by University of Craiova.

Institutional Review Board Statement: The study was conducted in accordance with the Declaration of Helsinki, and approved by the Ethics Committee of University of Craiova, University from Zilina and National R&D Institute for Welding and Material Testing—ISIM Timișoara

Informed Consent Statement: Not applicable

Data Availability Statement: The original contributions presented in the study are included in the article, further inquiries can be directed to the corresponding author.

Acknowledgments: The work performed is in line with the sustainability objectives of the project ANGIE, 2023-1-RO01-KA220-HED-00015803, co-financed by the European Commission through the program ERASMUS+. We would also like to thank the University of Craiova and the National R&D Institute for Welding and Material Testing for supporting this specific analysis.

Conflicts of Interest: The authors declare no conflicts of interest.

References

1. Project PN-III-P2-2.1-CI-2017-0088. *Optimizing the Constructive Elements of Electrical Protection and Switching Systems in Order to Improve the Life of the Product, Acronym REACTIV*; 2017. Available online: www.hipgnosis.ro/wp-content/uploads/2017/12/RST-Raport-Stiintific-si-Tehnic.pdf (accessed on 15 June 2024).
2. SERTIM Electric Annual Report. 2023. Available online: www.sertimelectric.ro (accessed on 20 December 2023).
3. Dwivedi, S.P.; Chaudhary, V.; Sharma, S. Synergistic performance evaluation of copper–WC-graphene composites: Microstructure, mechanical properties, corrosion, and wear behavior under microwave sintering. *Mater. Chem. Phys.* **2024**, *317*, 129202. [CrossRef]
4. Sun, Y.; Xu, L.; Zuo, C.; Tang, Z.; Xia, H.; Zheng, G. Fabrication of high thermal conductivity and wear resistance graphite/copper composites by microwave hot press sintering. *Tribol. Int.* **2024**, *194*, 109522. [CrossRef]
5. Wang, Z.; Tang, Z.; Xu, L.; Han, Z.; Liu, J.; Zhang, L. Thermal properties and thermal cycling stability of graphite/copper composite fabricated by microwave sintering. *J. Mater. Res. Technol.* **2022**, *20*, 1352–1363. [CrossRef]
6. Xu, L.; Srinivasakannan, C.; Zhang, L.; Yan, M.; Peng, J.; Xia, H.; Guo, S. Fabrication of tungsten–copper alloys by microwave hot pressing sintering. *J. Alloys Compd.* **2016**, *658*, 23–28. [CrossRef]
7. Kaushal, S.; Singh, S.; Bohra, S.; Gupta, D.; Jain, V.; Kapoor, M. Sustainable microwave processing and surface characterization of powdered tungsten reinforced copper metal matrix (Cu-Wx) castings. *Results Surf. Interfaces* **2024**, *16*, 100249. [CrossRef]

8. Cursaru, L.M.; Iota, M.; Piticescu, R.M.; Tarnita, D.; Savu, S.V.; Savu, I.D.; Dumitrescu, G.; Popescu, D.; Hertzog, R.G.; Calin, M. Hydroxyapatite from Natural Sources for Medical Applications. *Materials* **2022**, *15*, 5091. [CrossRef] [PubMed]
9. Stefan, I.; Savu, S.V.; Ghermec, O.; Nicolicescu, C.; Trotea, M. Research regarding the elaboration of barium hexaferrite from microwave heating. *Solid State Phenom.* **2012**, *188*, 369–375. [CrossRef]
10. Savu, I.D.; Savu, S.V.; Benga, G.C. Thermal runaway of the $BaCO_3$ + Fe_2O_3 homogenous mixture and mechanical alloys at the microwave heating. *Adv. Mater. Res.* **2014**, *837*, 185–189. [CrossRef]
11. Savu, S.V.; Tarnita, D.; Benga, G.C.; Dumitru, I.; Stefan, I.; Craciunoiu, N.; Olei, A.B.; Savu, I.D. Microwave technology using low energy concentrated beam for processing of solid waste materials from Rapana Thomasiana seashells. *Energies* **2021**, *14*, 6780. [CrossRef]
12. Vriezinga, C.A.; Pedreno-Sanchez, S.; Grasman, J. Thermal Runaway in microwave heating: A mathematical analysis. *Appl. Math. Model.* **2002**, *26*, 1029–1038. [CrossRef]
13. Menezes, R.R.; Souto, P.M.; Kiminami, R.H.G.A. Microwave fast sintering of ceramic materials. In *Sintering of Ceramics—New Emerging Techniques*; IntechOpen: London, UK, 2012; pp. 3–26.
14. Oliveira, A.V.S.; Teixeira, J.; Schick, V.; Gradeck, M.; Denis, S. Using a linear inverse heat conduction model to estimate boundary heat flux with material undergoing phase transformation. *Appl. Therm. Eng.* **2023**, *219*, 119406. [CrossRef]
15. Zhang, S.; Qin, S.; Wan, W.; Li, J.; Liu, S.; Paik, K.-W.; He, P.; Zhang, S. Investigation on electromagnetic and heat transfer coupling of heating susceptor for selective microwave hybrid soldering: 3D multi-physics simulation and experiment. *Appl. Therm. Eng.* **2024**, *225*, 124047. [CrossRef]
16. Said, M.; Salleh, N.A.; Nazeri, M.F.M.; Akbulut, H.; Kheawhom, S.; Mohamad, A.A. Microwave hybrid heating for lead-free solder: A review. *J. Mater. Res. Technol.* **2023**, *26*, 6220–6243. [CrossRef]
17. Savu, S.V.; Ghelsingher, C.D.; Stefan, I.; Sirbu, N.A.; Tarnita, D.; Simion, D.; Savu, I.D.; Bucse, I.G.; Tunescu, T. Microwave Soldering of Low-Resistance Conductive Joints-Technical and Economic Aspects. *Materials* **2023**, *16*, 3311. [CrossRef]
18. Savu, S.V.; Savu, I.D.; Benga, G.C.; Ciupitu, I. Improving functionality of Ti6Al4V by laser technology surfacing. *Optoelectron. Adv. Mater. Rapid Commun.* **2016**, *10*, 752–760.
19. Zhang, S.; Zhou, H.; Ding, T.; Long, W.; Zhong, S.; Paik, K.-W.; He, P.; Zhang, S. Grain orientations and failure mechanism of isothermal aged nano Sn-3Ag-0.5Cu/Cu solder joints by microwave hybrid heating and shear mechanical strength. *Eng. Fract. Mech.* **2024**, *2980*, 109902. [CrossRef]
20. Metaxas, A.C. Microwave heating. *Power Eng.* **1991**, *5*, 237–247. [CrossRef]
21. The Engineering Toolbox. Engineering Toolbox, Solids, Liquids and Gases—Thermal Conductivities. Available online: www.engineeringtoolbox.com (accessed on 10 June 2024).
22. The Engineering Toolbox. Engineering Toolbox, Specific Heat of common Substances. Available online: www.engineeringtoolbox.com (accessed on 15 June 2024).
23. Galaxy Global, Dolomite, dolomite.pdf. Available online: https://5.imimg.com/data5/PS/NE/SI/SELLER-105808956/dolomite-powder-100-mesh.pdf (accessed on 15 June 2024).
24. KAB Electro Acoustics, *Dielectric Constant Chart*, Dilectric-Constants.pdf. Available online: www.kabusa.com (accessed on 2 July 2024).
25. Garcia-Baños, B.; Catalá-Civera, J.M.; Peñaranda-Foix, F.L.; Plaza-González, P.; Llorens-Vallés, G. In Situ Monitoring of Microwave Processing of Materials at High Temperatures through Dielectric Properties Measurement. *Materials* **2016**, *9*, 349. [CrossRef]

materials

MDPI

Article

Influence of Laser Welding Modes along a Curved Path on the Mechanical Properties and Heterogeneity of the Microstructure of 316L Steel Plates

Dmitriy Andreevich Anufriyev [1], Vladimir Georgievich Protsenko [1], Maksim Vasilievich Larin [1], Mikhail Valerievich Kuznetsov [1,*], Aleksey Alekseevich Mukhin [2], Maksim Nikolaevich Sviridenko [2], Sergey Vyacheslavovich Kuryntsev [3,*], Oleg Ivanovich Grinin [4] and Yakov Borisovich Pevzner [4]

1 Institute of Machinery, Materials, and Transport, Peter the Great St. Petersburg Polytechnic University, Polytechnicheskaya, 29, Saint-Petersburg 195251, Russia; dmitriyanufriyev23@yandex.ru (D.A.A.); vova.protsenko1996@mail.ru (V.G.P.); xd.makc@gmail.com (M.V.L.)

2 N.A. Dollezhal Research and Development Institute of Power Engineering, Malaya Krasnoselskaya Str. 2/8, Moscow 107140, Russia; a.muhin@nikiet.ru (A.A.M.); sviridenko@nikiet.ru (M.N.S.)

3 Department of Materials Science and Welding, Kazan National Research Technical University Named after A.N. Tupolev—KAI (KNRTU—KAI), 10, K.Marx St., Kazan 420111, Russia

4 Institute of Laser and Welding Technologies, Saint Petersburg State Marine Technical University, Lotsmanskaya Str. 3, Saint Petersburg 190121, Russia; aspect@inbox.ru (O.I.G.); bezde@inbox.ru (Y.B.P.)

* Correspondence: kuznetsov_mich@mail.ru (M.V.K.); svkuryntsev@kai.ru (S.V.K.); Tel.: +7-911-237-16-46 (M.V.K.)

Abstract: The results of experimental studies in the manufacture of components of the supporting structure of the first wall panel, carried out as part of the manufacture of a model of the International Thermonuclear Experimental Reactor (ITER) using laser welding technology, are presented. The influence of laser welding modes on the quality of formation, microstructure characteristics, and mechanical properties of a welded joint made of 10 mm thick 316L steel was studied. A coaxial nozzle was designed and manufactured to protect the weld pool with a curved trajectory. The mechanical properties of the welded joint are 98–100% that of the base metal, and the microhardness of the welded joint and base metal is in the range of 180–230 HV. It was established that the lower part of the weld metal on the fusion line has transcrystalline grains and differs in δ-ferrite content; due to a high welding speed, the ratio of the depth to the width of the welding seam is 14 times. The width of the rectilinear part of the seam is 15–20% larger than its curved part.

Keywords: 316L stainless steel; high-speed welding; curved laser welding; microstructure; mechanical properties; transcrystalline grains

1. Introduction

High-alloy steels and alloys are the most important materials widely used in chemical, oil, power engineering, and other industries to manufacture the structures operating in a wide temperature range [1].

Currently, arc welding of large thicknesses (over 7–10 mm) has fewer and fewer advantages over laser welding. The main disadvantages of arc welding are high energy consumption, high heat input, significant overheating of the metal, and low speed of the welding process (about ten times) [2]. For steels of the austenitic class it is especially critical for multi-pass welding to withstand the interpass temperature, that is, to wait for the cooling of the weld metal of the previous pass of 150–170 °C. This significantly increases the duration of the technological process. The use of submerged arc welding will allow welding of large thicknesses (up to 20 mm) using one pass; however, this will lead to the formation of significant deformations and internal stresses in the weld metal and heat-affected zone (HAZ).

In this case, the use of welding with highly concentrated energy sources [3,4] is a promising solution to the problem. Welding with the application of high-quality high-power laser irradiation focused on a spot with a diameter of up to 50–150 μm is gaining special significance, as it allows welding of large thicknesses [5,6] with a depth (D) to width (W) ratio of 10–12 times. However, this topic has not been extensively studied, particularly in the case of high-speed welding (up to 100 mm/s).

During laser beam welding of thick workpieces, during the formation of a keyhole, in addition to ordinary defects specific defects such as root defects, middle cracks (longitudinal and transverse), humping and porosity also arise. In his works back in the 1990s, Katayama showed the reason for the formation of porosity, which, as a rule, is associated with metallurgical and hydrodynamic processes occurring during crystallization in the weld pool [7]. The main reason for hydrodynamic instability is the instability of the keyhole and plume behavior, justified by the Marangoni effect and the difference pressure in different parts of the keyhole [8]. The main methods of dealing with these defects are reducing the influence of harmful impurities (depending on the purity of the metal and shielding gas) and selecting the optimal welding modes [9].

Meanwhile, the appearance of middle cracks has a different nature and it is caused by the high tensile stresses and low plastic properties of the rapidly crystallizing weld metal. In this case, tensile stresses are directed both in the horizontal direction which leads to the formation of vertical cracks, and in the vertical direction which leads to the formation of horizontal cracks [10]. In the presented work, there was the formation of horizontal cracks in the middle of the weld, which was primarily caused by the large ratio of the weld depth to its width (10/1.5 mm). Such a large D/W ratio is inherent in electron beam welding [3,11].

Another reason for the formation of middle cracks is the metallurgical inhomogeneity of the weld metal. During the crystallization process of the weld pool, long columnar dendrites grow from the opposite walls of the unmelted metal, and push out low-melting eutectics at a temperature of 700–800 °C into the central part of the weld. This process is called dendritic segregation. Columnar crystallites grow continuously and do not change their direction until they collide with crystallites from the opposite direction [12]. This is especially true for high-speed welding. When welding with a defocused beam at a speed of 20 mm/s, crystallization of variously oriented dendrites is observed in the upper part of the weld [13]. In addition, in this case, due to the lower cooling rates, the formation of a more plastic δ-ferrite is provided, in which harmful impurities (sulfur and phosphorus) dissolve better.

On the other hand, in high-speed welding, there may not be enough time to separate and combine harmful impurities into separate clusters and their subsequent crystallization in the center of the weld [12,14].

According to the research results [15,16], a decrease in the heat input during a laser welding leads to a reduction in the length and width of the weld pool with an increase in spatter, leading to undercuts [16]. The authors of [17] described the research results of the influence of heat input and the location of the focal plane relative to the surface to be welded on the quality of the welded joint formation. It is noted that an increase in the heat input, as well as the focal plane position above the surface to be welded, leads to the occurrence of undercuts and sagging of the weld root in a particular range of welding speeds. With the deepening of the focal plane, the absence of undercuts and the presence of a uniform welded joint with high-quality formation with an increased depth of penetration were noted.

In [5,18], the experimental study results of the influence of heat input on the characteristics of a keyhole are presented. An increase in the heat input leads to an increase in the stability and diameter of the keyhole; however, at the same time, large cavities are formed in the bottom of the keyhole, which actively migrate into the melt pool, with the subsequent formation of pores. The research [19] also revealed that the power of laser irradiation has more effect on the geometry of the welding bead compared to the welding speed. According to the study results, the microstructure of the weld metal formed with a lower heat input contains less δ-ferrite in the absence of external and internal defects, such as porosity and cracks, and the welded joint has a desired quality level and meets the requirements of ISO 13919 [20].

The works [5,21] describe the study results of the influence of shielding gas on the geometry of the weld during laser welding. It is noted that the weld seam when using nitrogen as a shielding gas is deeper than when using argon. Also, the presence of nitrogen led to a narrowing of the weld root and expansion of the top part of the keyhole.

In [22,23], similar values of hardness were obtained in the region of the weld metal, and in the fusion zone, an increase in the value of hardness is observed due to the refinement of the grain size. According to the research results [23,24], preliminary application of chromium oxide (Cr_2O_3) to the welded surface increases the stability of the welding process, prevents the formation of pores, reduces spatter, stabilizes the keyhole behavior, refines the grain, narrows the width of the HAZ and increases the strength characteristics of the welded joint.

Also, studies that explore the use of a hybrid laser-arc method of welding sheets of large thicknesses from stainless steel in several passes were carried out [6,25]. The researchers found that the metal of the remelted interpass areas after reheating has significantly different grain size and orientation, which leads to the formation of residual stress, which is also confirmed by the simulation results.

Analysis of recently published articles (Table 1) shows that high-speed laser welding (up to 100 mm/s) for products of large thickness (>10 mm) is not widely used. Works on high-speed laser welding (200–800 mm/s) of small and medium thicknesses (0.3–3 mm) have been published mainly for welding dissimilar metals or for developing the methods of industrial monitoring of high-speed laser welding [14,26,27]. Scientific publications studying the microstructure and mechanical properties of the weld metal obtained by laser welding along a curved trajectory were not found. Based on the above, the use of high-speed laser welding along a curved trajectory of austenitic steels and other alloys is a promising direction for research and development of technologies.

Also, the influence of welding speeds and, accordingly, cooling rates leads to the formation of δ-ferrite and microstructure heterogeneity, which affects the corrosion resistance of the weld metal [28–30]. However, the δ-ferrite content (5–15%) in weld metal has a positive effect on resistance to the formation of hot cracks and the dissolution of harmful impurities (sulfur and phosphorus).

Table 1. Analysis of welding modes.

№	Reference	S, mm	Power, kW	Welding Speed, mm/s	D_{spot}, mm	Shielding Gas, L/min	Steel Grade	Laser Source
1	[15,16]	0.6	1.5 2.5 3.5	30 50 70	-	Ar, 15 He, 10	AISI 316L	CO_2 Nd: YAG
2	[30,31]	5	4	20	-	Ar, 10	AISI 316L	-
3	[16,17]	2.7	2.8	16.6	1.2	Ar, 80	AISI 316L	Nd: YAG
4	[23,24]	5.6	2.5	16.6	-	He, 20	AISI 316L(N)	CO_2
5	[19,20]	5	1.5; 2 2	6.1 13.3	0.6	Ar, 30	AISI 316L	Nd: YAG
6	[22]	-	2.6	10	-	He	AISI 316L	CO_2
7	[21]	Up to 18	Up to 10 kW	70–90	0.13 0.2 0.36 0.56	Ar, 30 N_2, 30	AISI 304	fiber laser

Table 1. *Cont.*

№	Reference	S, mm	Power, kW	Welding Speed, mm/s	D_{spot}, mm	Shielding Gas, L/min	Steel Grade	Laser Source
8	[5]	11	6	10, 50, 100, 166	0.13	Ar	AISI 304	fiber laser
9	[13,17]	12 10	10	15–45	0.4	Ar, 30 N_2, 30 He, 30	AISI 321	fiber laser

Also, the influence of welding speeds and, accordingly, cooling rates leads to the formation of δ-ferrite and microstructure heterogeneity, which affects the corrosion resistance of the weld metal [27–29]. However, the δ-ferrite content (5–15%) in weld metal has a positive effect on resistance to the formation of hot cracks and the dissolution of harmful impurities (sulfur and phosphorus).

The paper presents a study of the influence of the high-speed laser welding parameters along a curved contour (HSCLW) of 316L grade austenitic steel on the geometry, microstructure, and mechanical properties of the welded joint. This study aims to select the optimal laser welding modes, to carry out metallographic studies and mechanical tests that establish the applicability of high-speed laser welding along a curved trajectory for large thicknesses, and to establish the effect of high cooling rates on the heterogeneity of weld metal in depth.

2. Experimental

The objectives of this study are to determine the influence of mode parameters on the quality of formation and mechanical characteristics of the welded joint during laser welding of 316L steel with a thickness of 10 mm. Moreover, the laser welding technology that will ensure the formation of a welded joint that meets the requirements of quality level B according to ISO 13919 [20] is being developed. Within the framework of the experimental study, such parameters of laser welding as the power of laser radiation, welding speed at a constant heat input, the focus depth relative to the welded surface, and the quantitative and qualitative composition of the shielding gas were varied. A one-factor experiment was carried out to assess the influence of each mode parameter on the studied characteristics.

2.1. Experimental Equipment and Welding Materials

The experiments were carried out in the Laboratory of Laser and Additive Technologies, St. Petersburg State Marine Technical University(Russia), using a portal-type technological laser complex. Figure 1 shows the appearance of the complex. Table 2 gives the technical characteristics of the complex.

The complex is based on a YLS-16 ytterbium fiber laser (IPG) with a maximum output power of 16 kW, linear guides (Isel), and a HighYAG Bimo HP laser head, with focal lengths of the focusing and collimating lenses of 300 mm and 150 mm, respectively, rigidly fixed on the carriage which is moved along the linear guides of the portal. Laser irradiation was supplied to the laser head through a fiber of 100 μm in diameter. In this case, a focal spot of 200 μm was used, and the small width was also affected by the low thermal conductivity of this steel.

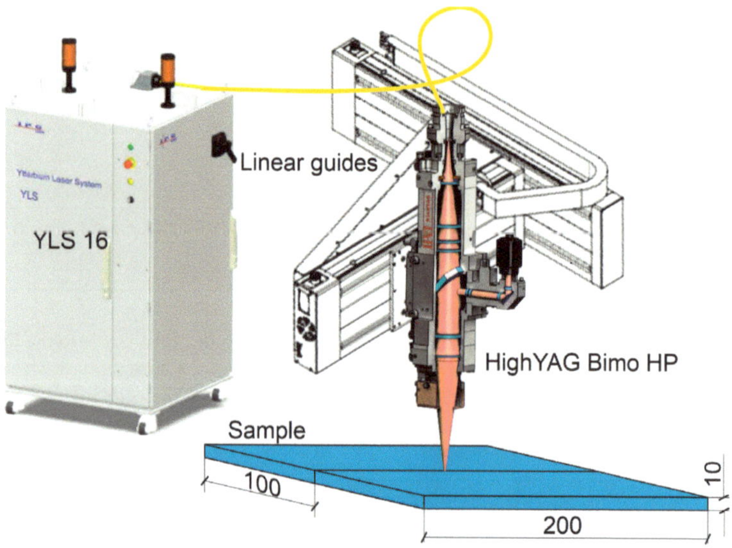

Figure 1. External view of the technological laser complex.

Table 2. Technical characteristics of the technological complex.

№	Parameter, Unit of Measurement	Value
1	Maximum power of the laser source, kW	16
2	Maximum linear speed, m/min	4
3	Adjustment range of linear movement speed, m/min	from 0.4 to 4
4	Maximum speed of idle movements, m/min	10
5	Type of movement control	CNC
6	Maximum deviation from straightness when moving, mm	±0.5

Two sets of 316L steel samples were prepared for testing and certification of the laser welding technology. The first set of samples with overall dimensions of 200 × 100 × 10 mm was used for testing the technology, while the second set of samples with overall dimensions of 450 × 150 × 10 mm was used for certification. The samples were cut using plasma cutting from 316L steel, after which the welded edges were milled. Before welding, the surfaces mating with the welded edge were polished to a metallic sheen, using an angle grinder; the welded edges of the samples were then degreased with acetone. The tacks were placed on the reverse side of the surface to be welded with a step between the tacks of 100 mm and 50 mm from the end of the plate. The shielding gas used was top-grade inert argon gas, according to the requirements of ISO 14175-I1 [32], with a flow rate of 50 L/min on the top side and 25 L/min on the bottom side. A nozzle with a diameter of 100 mm was designed and manufactured to shield the weld pool from the top side during laser welding along a curved path, which provides coaxial protection of the melt pool and crystallized metal of the welded joint (Figure 2),which is heated to a temperature above 200 °C. A device to ensure uniform distribution of the shielding gas along the entire length of the welded joint was made to protect the bottom of the weld.

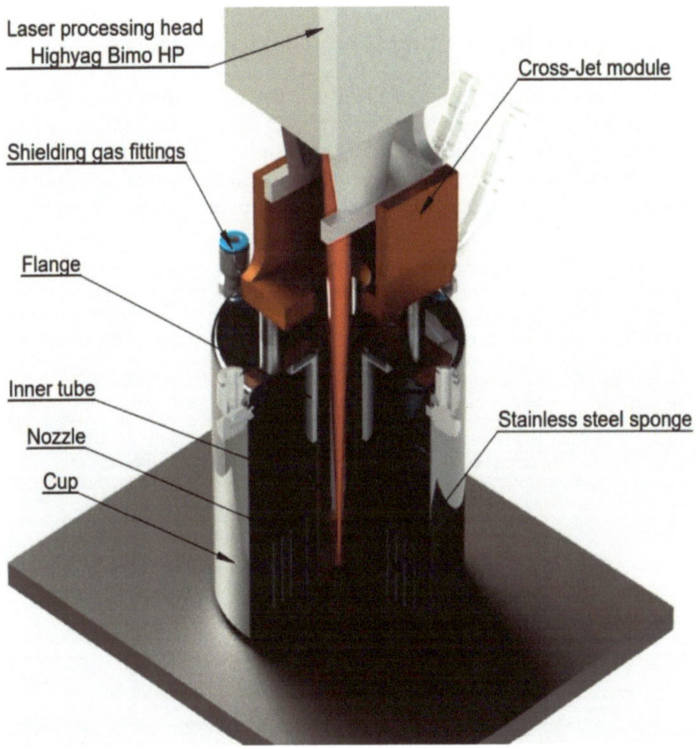

Figure 2. Nozzle for coaxial protection of the melt pool.

In this work, AISI 316L steel was used—an improved version of 304 steel, with the addition of molybdenum and a higher nickel content. This chemical composition of AISI 316L significantly improves corrosion resistance in most aggressive environments. Molybdenum provides protection against pitting and crevice corrosion in chloride environments, seawater, and acetic acid vapors.

AISI 316L steel has higher strength and better creep resistance at high temperatures than AISI 304. AISI 316L steel also has excellent mechanical properties and corrosive characteristics at low temperatures. Table 3 presents the chemical composition, physical and mechanical properties of 316L steel.

Table 3. Chemical composition of 316L steel (mass fraction: %).

	C	Si	Mn	Ni	Cr	Mo	Fe	S	P
Base metal	0.03	0.75	2	12	16	2.5–3	~67	0.030	0.045

2.2. *Laser Welding Procedure*

The following main variable parameters of the laser welding mode were chosen:

- Power of laser irradiation in the range from 7 to 15 kW;
- Welding speed in the range from 30 to 75 mm/s;
- Depth of the focal plane of the focusing lens relative to the surface in the range from 0 to 10 mm;
- Tilt angle of the laser beam;
- Shielding gas composition: Ar, N_2.

The ranges of varying parameters were chosen on the basis of analyzing the works of other authors. Table 4 shows the parameters of welding modes with the highest quality of weld formation on the front and bottom sides during laser welding along a straight path. This table shows 15 variants of samples, while in reality more than 40 variants of various laser welding modes were processed. As can be seen from Table 5, the heat input was in the range of 182.5–320 W × s/mm. The parameters of the modes in Table 4 were obtained using an iterative method based on visual control of welded specimens. The selection criterion was the absence of external defects (cracks, undercuts, craters, and pores).

Table 4. Modes of laser welding.

№	Power, W	Welding Speed, mm/s	ΔF, mm	Tilt Angle, ○	Shielding Gas	Shielding Gas Flow Rate, L/min	Heat Input, W × s/mm
1	14.00	75	−5	0	Ar	50	193.3
2	12,800	40	−10	10	Ar	50	320.0
3	9700	40	−10	10	Ar	50	242.5
4	12.800	40	−1	10	Ar	50	320.0
5	7300	40	−1	5,5	Ar	50	182.5
6	7660	40	−5	0	Ar	50	191.5
7	7761	30	−5	0	Ar	50	258.7
8	13.020	60	−5	0	Ar	50	217.0
9	15.000	65	0	0	Ar	50	230.8
10	15.000	70	0	0	Ar	50	214.3
11	15.000	75	0	0	Ar	50	200.0
12	7640	30	0	0	Ar	50	254.7
13	15,000	75	0	0	N_2	50	200.0
14	7761	30	−5	0	N_2	50	258.7
15	7761	32.5	0	0	N_2	50	238.8

Table 5. Calculation of the ratio of the area of defects to the area of the weld.

Parameter	Value
Width, mm	1.8
Length, mm	130
Area, mm^2	234
Number of defects	1
Defect diameter, mm	0.3
Defect area, mm^2	0.07
Total area of defects, mm^2	0.07
Area ratio of defects to seam,%	0.03

2.3. Research Equipment and Quality Control Methods for Welded Joints

The quality of the welded joint formation was assessed using optical and visual measuring methods of control. The optical method was implemented using a Leica DMi8 A optical microscope (Leica, Germany), according to BS EN 61326-1:2013 [33]. Visual measurement control (BS EN ISO 23277:2015 [34]) was carried out according to the requirements of ISO 13919-1 (quality level B) [20].

Non-destructive testing of welded joints was performed using the radiographic method on an RPD-250SP X-ray machine («Nerazrushayushchiy control», Yekaterinburg, Russia) with a control sensitivity of 200 µm. Welding requirements were according to ISO 13919-1 (quality level B).

For certification of welding procedures according to the requirements of ISO 15614-11 [35] and qualification of fusion welding operators according to the requirements of BS EN ISO 14732 [35], 2 samples were welded for subsequent testing in a certified laboratory.

Welded joints were subjected to capillary inspection according to ISO 3452-1 [36] and radiographic inspection according to ISO 17636-1 [37]. The non-destructive testing was carried out by certified level-2 technicians in accordance with ISO 9712 [38]. The VIC KVK-A kit (SRTC WT&NDT «SPEKTR», Moscow, Russia) was used. The results of non-destructive testing were documented and confirmed by the protocol.

Also, samples made from welded joints were subjected to destructive testing: the tensile test (EN ISO 4136 [38]), bending test (EN ISO 5173 [39]), Charpy test (EN ISO 9016 [40]), and hardness measurements (EN ISO 9015-1 [41]). The destructive test methods were carried out in a laboratory accredited according to ISO 17025 [42] at room temperature. Static bending and tensile testing of the weld was carried out according to the ISO 5143 [43] loading rate of 10 mm/min. The tests were carried out on a universal testing machine LFM-250 kN (Walter + Bai AG, Löhningen, Switzerland).

The study of the structure was carried out on microsections according to EN ISO 17639 [44]. An intergranular corrosion test was conducted according to EN ISO 3651-1 [45] method 2. The tests were carried out on a universal testing machine LFM-250 kN (Walter + Bai AG, Löhningen, Switzerland) and a complex of hardware and software analysis Thixomet (Thixomet, St.Petersburg, Russia). Samples were etched with a solution of HCl with HNO_3 in a ratio of 3:1.

Weld microhardness was measured with a Remet HX-1000 microhardness tester (Remet S.A.S., Casalecchio di Reno, Italy) using a load of 0.1 kg and a dwell time of 10 s, according to DIN EN ISO 14577 [46], at room temperature.

3. Results

3.1. Appearance and X-ray Inspection

Samples with both straight welds and curved welds were subjected to capillary inspection and radiographic inspection (Figure 3).

Figure 3. Appearance of the weld obtained in mode12; (**a**) top side, (**b**) bottom side, (**c**) X-ray.

Digital X-ray images of several samples obtained using mode 1 were analyzed using the Digimiser 6.3.0 software (MedCalc Software, Ostend, Belgium). The linear dimensions of the internal defects of the weld, namely, pores and slag inclusions, were obtained. Next, the total projection area of the weld defects was calculated, and the ratio of the projection area of the weld defects to that of the weld was obtained. According to ISO 13919-1, this ratio must be less than 0.7%.

Figure 4 presents photographs of the appearance of the samples and radiographs of samples welded along a curved trajectory. It shows that no defects are observed in the weld areas with a curvilinear trajectory.

Figure 4. Photographs of the appearance of the samples and radiographs of samples welded along a curved trajectory (mode 1).

3.2. Mechanical Tests

Weld impact tests were carried out in accordance with ISO 9016. The location of the notch is parallel to the symmetry axis of the seam. Tests were carried out on six specimens welded on mode №1. The test results are presented in Table 6.

Table 6. Test results of the weld impact test.

Sample№	Test Temperature, °C	Notch Location	Impact Energy, J	Impact Strength, J/cm^2	Average Impact Test, J/cm^2
1	+20	Middle of the welding seam	88.2	147	168.44 (Base metal 160–180)
2	+20		110.5	184.17	
3	+20		101.1	168.5	
4	+20		96	160	
5	+20		102.7	171.17	
6	+20		107.9	179.83	

Static tensile testing of the weld was carried out in accordance with ISO 4136. Loading speed was 8 mm/min. The tests were carried out using a universal testing machine LFM-250 kN. The test results are presented in Table 7.

Table 7. Results of tensile strength test of a welded joint.

Sample №	Sample Cross-Sectional Square, mm^2	Tensile Force, kN	Tensile Strength, MPa	Fracture Location
1	253.28	142.84	563.96 (570 base metal)	Base metal
2	246.31	140.35	569.8 (570 base metal)	Base metal

Static three-point bending tests of the weld were carried out according to ISO 5143. Loading speed was 10 mm/min. The tests were carried out on a universal testing machine LFM-250 kN. The test results are presented in Table 8.

Table 8. Results of three-point bending test.

Sample №	Bending Location	Bending Angle, °Deg.	Defects
1	Top	180	No defects
2		180	No defects
3	Bottom	180	No defects
4		180	No defects

Additionally, the tests were carried out on the resistance of the welded joint to intergranular corrosion. The tests were carried out according to ISO 3651-2 [47] method 2. Based on the test results, it was revealed that the samples of the welded joint do not show a tendency to intergranular corrosion.

The measurement of microhardness values was carried out across the weld on its upper and lower parts, as usual (black lines, Figure 5). Separately, measurements of the microhardness of the upper surface (red line, Figure 5) and several measurements in the middle part at different depths (green line, Figure 5) were carried out. In general, the microhardness of the entire weld was in the range of 170–230 HV. The maximum microhardness was observed in the middle part of the red line, that is, at the top of the bead 230–235 HV and in the base metal 220–230 HV. It may be due to the higher cooling rate at the top of the bead. The minimum microhardness was observed at the fusion line in the upper part of the "Martini glass" shape of the weld (Figure 5b), at 170–185 HV. This may be due to the large volume of melt in this zone. In the middle part, along all green lines, the microhardness is in the range of 195–210 HV, and the maximum microhardness is in the center of the weld, where crystals growing from opposite fronts grow together. The range of the root part of the weld 170–190 HV is slightly lower. Also, in the root part of the weld, different microhardness was measured in the volume of the same grain, part of which was in the base metal (197 HV), and the other part was behind the fusion line, in the weld (186 HV) (Figure 5e).

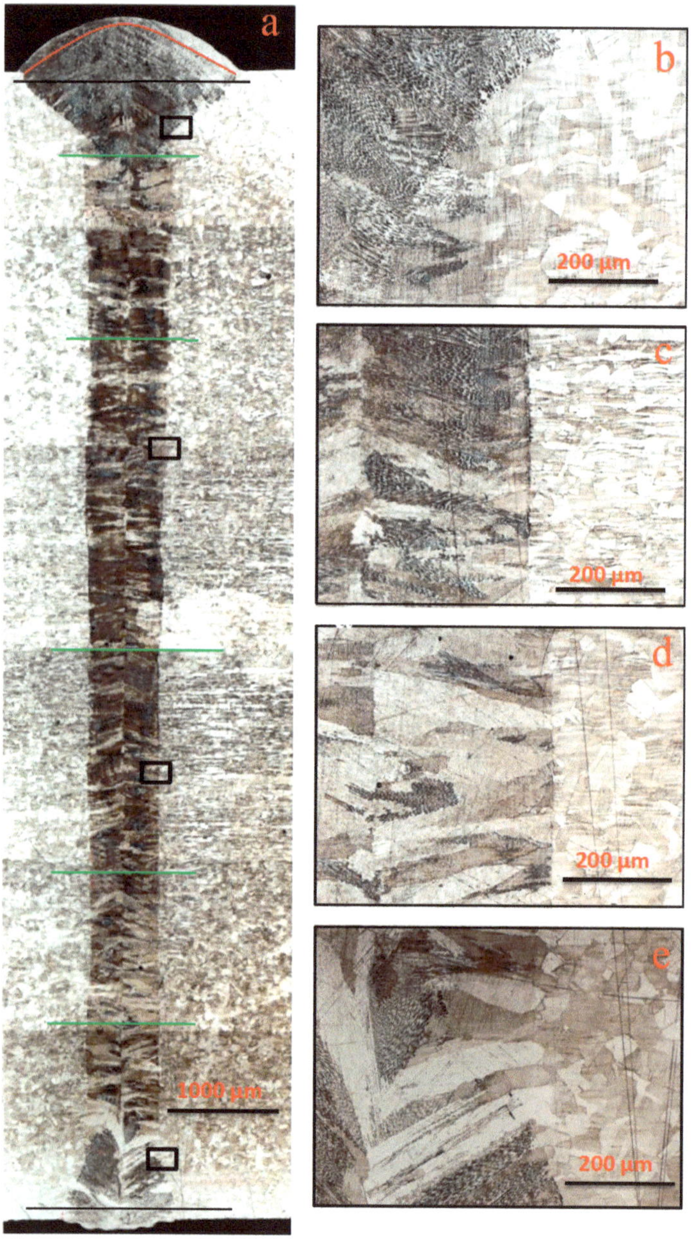

Figure 5. Macro- and microstructure of the weld obtained according to mode 1. (Q = 193.3 W × s/mm). Whole weld (**a**), upper part of the weld (**b**), middle part of the weld at different depths (**c**,**d**), lower part of the weld (**e**).

3.3. Metallographic Studies

3.3.1. Macrostructure

For metallographic studies, macrosections of the longitudinal section of the welded samples from 316L steel 10 mm thick and the cross section from the top surface of the weld were made (Figure 6). On the macrostructure photographs of the top surface of the weld, the width of the weld pool was measured (50 measurements). The dimensions of the weld pool width were in the range of 1132–1733 microns; smaller values of the width corresponded to the rectilinear part of the weld and bigger ones to the curved part. In some areas, an alternate increase and decrease in width was observed in the range of 50–70 microns, which is caused by fluctuations in the weld pool, despite very high welding speeds. The waviness of the fusion line along the inner radius is more frequent compared to the outer radius, which indicates the asymmetry of the weld pool behavior and requires more detailed study in the future.

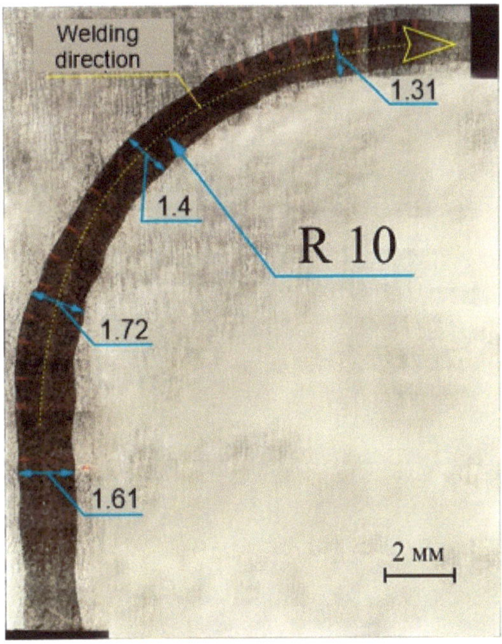

Figure 6. Photo of the macrostructure of the top transverse surface of the weld.

Figure 7 shows photographs of the macrostructure of welded joints selected from those welded according to the modes presented in Table 4, and the corresponding heat input. The heat input of the selected samples is in the range of 182.5–258.7 W × s/mm. It can be seen that the samples welded using the same heat input (7 and 14 258.7 W × s/mm) and speed and power, but using different shielding gases, differ significantly. Samples 1 (193.3 W × s/mm) and 6 (191.5 W × s/mm) welded with almost the same heat input, but with two-times different speed (1–75 mm/s, 6–40 mm/s) and power (1–14,500 W, 6–7660 W), do not differ significantly in macrostructure. However, a high welding speed helps to increase productivity and minimize thermal deformations. Samples welded according to modes 3, 5, 9, 13, and 14 have a lack of penetration or significant defects in the root part of the welding seam. Samples 10 and 11 have defects in the form of porosity and weld root undercut, although, in general, these defects are not so significant.

The macrostructure of the weld looks like a "Martini glass" shape. The width of the upper part is 1938 μm, the main width is in the range of 625–760 μm, and the width of the root part is 1025 μm. A main width-to-depth ratio is D/W_{av} = 14. At a depth of 90% of the weld, the opposite sides are parallel, which has a positive effect in terms of internal tensile stresses. Since welding was carried out without filler material and without a gap due to a change in the phase composition and the formation of δ-ferrite, the volume of the weld metal increased. As a result, the lower and upper beads were formed, with a total volume of 0.745 mm^2, while the volume of the rest of the weld, not extending beyond the top and bottom surfaces, was 7.46 mm^2. This shows that because of phase transformations of austenite into a mixture of austenite and δ-ferrite, the metal volume increased by about 10%.

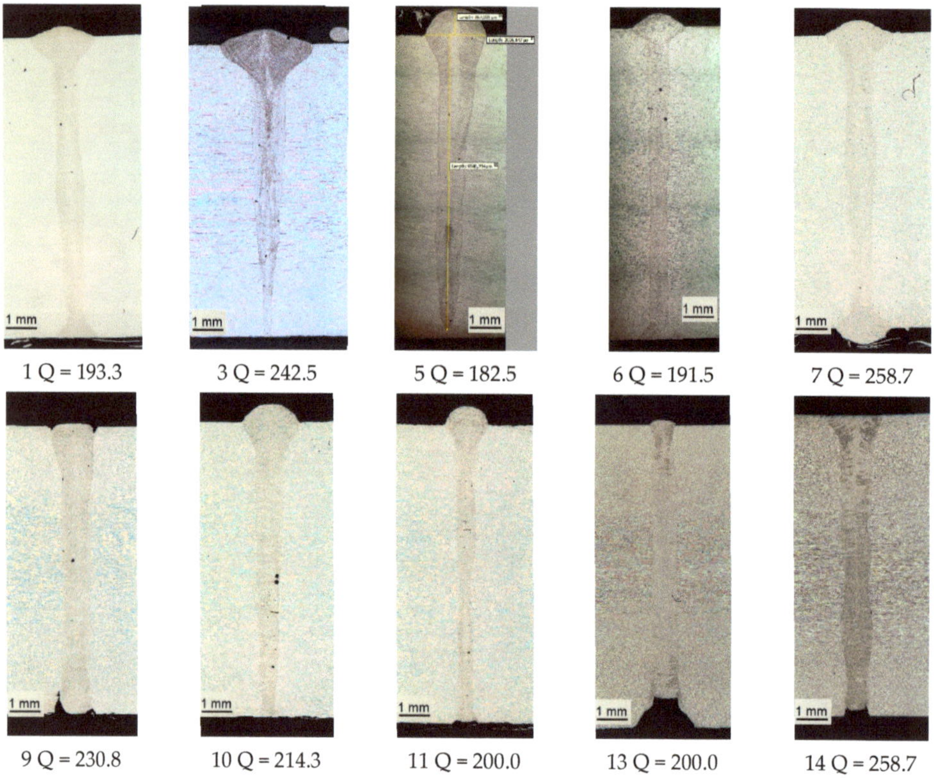

Figure 7. Macrostructure of welds obtained using different heat inputs.

3.3.2. Microstructure

The microstructure of the fusion line of the upper and lower parts of the weld differs in composition and nature, as in the previous work of the authors [29,30]. As can be seen in Figure 5, the fusion line of the upper and upper-middle parts of the weld is different from the lower and lower-middle parts of the weld. Figure 5b,c show that the transition across the fusion line is abrupt from the base metal structure with coarse austenite grains to the cast dendritic structure of the weld with a large amount of δ-ferrite, whereas in Figure 5d,e, the fusion line is not so abrupt, and contains transcrystalline grains that continue to grow from the base metal through the fusion line into the weld metal. This is clearly seen when comparing microstructure images of the seam root part, Figure 8a,b. A fusion line with an abrupt transition is observed at about 60% of the weld depth from the top. On the remaining 40%, a fusion line with a transcrystalline structure is observed. Most likely, transcrystallization occurs as a result of lower cooling rates of the middle and bottom parts of the weld compared to the upper part, as well as a lesser degree of overheating of the middle and bottom parts of the seam. Although when welding with a laser beam the energy source is supplied from the top side, the middle and bottom parts of the weld remain in a heated and molten state for a longer time. A similar effect has already been seen in [31]. Microhardness measurements were carried out with a load of 25 g, after which the imprint allowed for the making of several measurements inside one grain to study the properties of the transcrystalline grains. The obtained values were within the range of 185–192 HV and were practically identical.

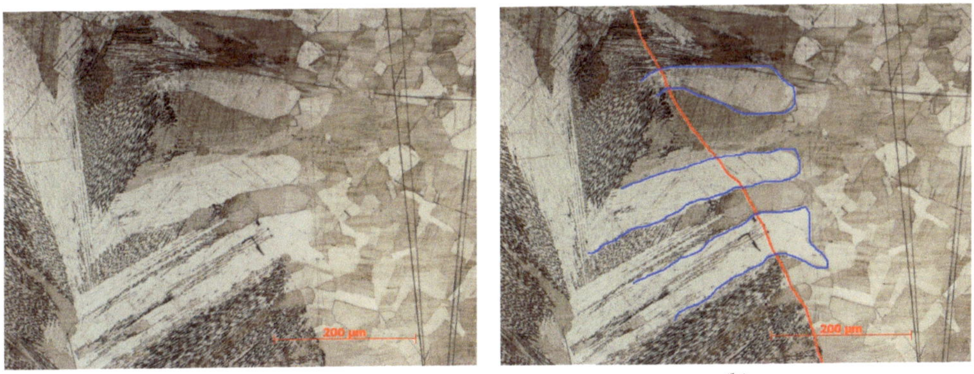

(**a**) (**b**)

Figure 8. Microstructure of the lower part of the weld, (**a**) without allocation of transcrystalline grains, (**b**) with allocation of transcrystalline grains (blue color) and fusion line (red color).

In the previous work of the authors [48], it was found that during multi-pass welding, crystallization in the austenite–ferrite solidification mode was observed in the center of the middle pass, while, according to the calculation of Cr_{eq}/Ni_{eq} = 1.62, only the ferrite–austenite solidification mode should be observed. In this case, no such effect is observed, and the central fusion line is very clear (Figure 5d). Also, in comparison with previous work, the fusion line with the base metal is significantly different, and is abrupt (Figure 5b), especially (Figure 5c,d), and in the weld root there are only transcrystalline grains (Figure 5e). In [48], a large amount of columnar δ-ferrite is observed at the fusion lines with the base metal, which is significantly different from the microstructure of the HSCLW seam (Figure 5). This indicates a significant influence of cooling rates on the phase composition and microstructure of the weld metal.

The maximum amount of δ-ferrite is observed at the top and bottom of the weld, especially at the fusion line, where high cooling rates reach maximum values (Figure 5b,e). There are no columnar δ-ferrite grains on the fusion line of the middle part of the weld (Figure 5d). In all parts of the weld, a mixed columnar and skeletal δ-ferrite structure is observed. The growth direction of δ-ferrite crystallites differs significantly, depending on the part of the weld: in the upper part of the weld +45°; in the middle part from 0° to −35° relative to the horizontal.

These observations show a significant influence of HSCLW on the heterogeneity of the weld metal microstructure in different areas; however, this does not lead to negative consequences such as the formation of defects or a decrease in mechanical properties. In the future, it would be beneficial to conduct a comprehensive quantitative study of the content and structure of δ-ferrite and to study the microstructure of the top and bottom parts of the weld in the curved section of the weld.

4. Conclusions

High-speed single-pass laser welding was used to join 10 mm thick 316L steel workpieces. Optimal laser welding modes were selected experimentally. Studies of the influence of optimal welding conditions on the heterogeneity of the microstructure, mechanical properties, and corrosion resistance of the welded joint were carried out.

1. A device that allows high-quality protection of the top of the welding pool (before it is cooled below 200 °C) during high-speed laser welding along a curved path was designed and developed.

2. The mechanical properties of the welded joints are 98–100% of that of the base metal. Corrosion resistance complies with EN ISO 3651-1 standard. The microhardness of the weld metal does not differ significantly from the base metal (180–230 HV). Along the depth of the weld, the microhardness is non-uniform; the maximum microhardness was observed in the middle part of the upper bead 230–235 HV, and the minimum microhardness was observed on the fusion line in the upper part of the weld 170–185 HV, which is associated with a significant difference in cooling rates.
3. The main features of a weld formed using HSCLW is the ratio of depth to average width (D/W) times 14; a fusion line with an abrupt transition is observed at 60% of the weld depth from the top and on the remaining 40% a fusion line with a transcrystalline structure is observed, which depends on the welding thermal cycle. A change in the weld trajectory affects the geometric dimensions of the weld pool; the width of the curved part of the weld is 15–20% less than the width of the rectilinear part.
4. The high speed of laser welding (up to 4.5 m/min) has a positive effect on the cross-sectional shape of the seam; the main width of the seam is in the range of 625–760 microns, which for a welded workpiece thickness of 10 mm is very small, has a beneficial effect on the uniform heating of the metal across the thickness, and helps to minimize residual stresses.

Author Contributions: Conceptualization, project administration, funding acquisition, M.V.K., M.N.S., A.A.M. and Y.B.P.; methodology, resources A.A.M., D.A.A., M.V.L., Y.B.P. and O.I.G.; validation, investigation, data curation, V.G.P., D.A.A., M.N.S., O.I.G. and S.V.K.; writing—original draft preparation, writing—review and editing, M.V.L., V.G.P., M.V.K. and S.V.K. All authors have read and agreed to the published version of the manuscript.

Funding: The research is partially funded by the Ministry of Science and Higher Education of the Russian Federation as part of the World-class Research Center program: Advanced Digital Technologies (contract No. 075-15-2022-311 dated April 20, 2022).

Institutional Review Board Statement: Not applicable.

Informed Consent Statement: Not applicable.

Data Availability Statement: The data presented in this study are available on request from the corresponding author.

Conflicts of Interest: The authors declare no conflict of interest.

References

1. Touileb, K.; Attia, E.; Djoudjou, R.; Benselama, A.; Ibrahim, A.; Boubaker, S.; Ponnore, J.; Ahmed, M.M.Z. Laser Weld Aspect Optimization of Thin AISI 316 SS Using RSM in Relation with Welding Parameters and Sulfur Content. *Metals* **2023**, *13*, 1202. [CrossRef]
2. Liao, H.; Zhang, W.; Xie, H.; Li, X.; Zhang, Q.; Wu, X.; Tian, J.; Wang, Z. Effects of welding speed on welding process stability, microstructure and mechanical performance of SUS304 welded by local dry underwater pulsed MIG. *J. Manuf. Process.* **2023**, *88*, 84–96. [CrossRef]
3. Verma, J.; Taiwade, R.V. Effect of welding processes and conditions on the microstructure, mechanical properties and corrosion resistance of duplex stainless steel weldments—A review. *J. Manuf. Process.* **2017**, *25*, 134–152. [CrossRef]
4. Sokolov, M.; Salminen, A.; Kuznetsov, M.; Tsibulskiy, I. Laser welding and weld hardness analysis of thick section S355 structural steel. *Mater. Des.* **2011**, *32*, 5127–5131. [CrossRef]
5. Kawahito, Y.; Mizutani, M.; Katayama, S. Elucidation of high-power fibre laser welding phenomena of stainless steel and effect of factors on weld geometry. *J. Phys. D Appl. Phys.* **2007**, *40*, 5854–5859. [CrossRef]
6. Xu, Z.; Wang, J.; Yan, C.; Ren, J.; Zhou, Y.; Li, Y.; Zhan, X. Inhomogeneity of microstructure and mechanical properties in the interlayer regions for narrow gap laser wire filling welding of 316L stainless steel. *Opt. Laser Technol.* **2024**, *169*, 110050. [CrossRef]
7. Seto, N.; Katayama, S.; Matsunawa, A. Porosity formation mechanism and reduction method in CO_2 laser welding of stainless steel. *Weld. Int.* **2002**, *16*, 451–460. [CrossRef]
8. Fuhrich, T.; Berger, P.; Hügel, H. Marangoni effect in laser deep penetration welding of steel. *J. Laser Appl.* **2001**, *13*, 178–186. [CrossRef]
9. Lippold, J.; Kotecki, D. *Welding Metallurgy and Weldability of Stainless Steels*; Wiley-VCH: Hoboken, NJ, USA, 2005.

10. Shiganov, I.N.; Grezev, N.V.; Shamov, E.M. Special features of welding 10KhSND low-alloy structural steel with an oscillating laser beam. *Weld. Int.* **2017**, *31*, 733–738. [CrossRef]
11. Chen, Y.; Liu, G.-M.; Li, H.-Y.; Zhang, X.-M.; Ding, H. Microstructure, strain hardening behavior, segregation and corrosion resistance of an electron beam welded thick high-Mn TWIP steel plate. *J. Mater. Res. Technol.* **2023**, *25*, 1105–1114. [CrossRef]
12. Kadoi, K.; Fujinaga, A.; Yamamoto, M.; Shinozaki, K. The effect of welding conditions on solidification cracking susceptibility of type 310S stainless steel during laser welding using an in-situ observation technique. *Weld. World* **2013**, *57*, 383–390. [CrossRef]
13. Kuryntsev, S.V. Effect of Heat Treatment on the Phase Composition and Corrosion Resistance of 321 SS Welded Joints Produced by a Defocused Laser Beam. *Materials* **2019**, *12*, 3720. [CrossRef]
14. Ai, Y.; Jiang, P.; Wang, C.; Mi, G.; Geng, S.; Liu, W.; Han, C. Investigation of the humping formation in the high power and high speed laser welding. *Opt. Lasers Eng.* **2018**, *107*, 102–111. [CrossRef]
15. Yan, S.; Shi, Y.; Liu, J.; Ni, C. Effect of laser mode on microstructure and corrosion resistance of 316L stainless steel weld joint. *Opt. Laser Technol.* **2019**, *113*, 428–436. [CrossRef]
16. Tanigawa, H.; Aburadani, A.; Shigematsu, S.; Takeda, N.; Kakudate, S.; Mori, S.; Jokinen, T.; Merola, M. Comparative study of laser and TIG welding for application to ITER blanket hydraulic connection. *Fusion Eng. Des.* **2012**, *87*, 999–1003. [CrossRef]
17. Zhang, M.; Chen, G.; Zhou, Y.; Liao, S. Optimization of deep penetration laser welding of thick stainless steel with a 10kW fiber laser. *Mater. Des.* **2014**, *53*, 568–576. [CrossRef]
18. Zhengwu, Z.; Xiuquan, M.; Chunming, W.; Gaoyang, M. Grain refinement and orientation alternation of 10 mm 316L welds prepared by magnetic field assisted narrow gap laser-MIG hybrid welding. *Mater. Charact.* **2020**, *164*, 110311. [CrossRef]
19. Arivarasu, M.; Vishnu, G.; Hari, P.; Vipin, V.; Katiki, K.; Gokulkumar, K.; Manikandan, M.; Ramkumar, K.D.; Arivazhagan, N. Micro-segregation Studies on the Continuous Nd: YAG Laser Beam Welded AISI 316L. *Procedia Eng.* **2014**, *97*, 892–901. [CrossRef]
20. ISO 13919-1:2019 [Electronic Resource]//ISO. Available online: https://www.iso.org/ru/standard/75514.html (accessed on 20 November 2023).
21. Kawahito, Y.; Mizutani, M.; Katayama, S. High quality welding of stainless steel with 10 kW high power fibre laser. *Sci. Technol. Weld. Join.* **2009**, *14*, 288–294. [CrossRef]
22. Tjong, S.; Zhu, S.; Ho, N.; Ku, J. Microstructural characteristics and creep rupture behavior of electron beam and laser welded AISI 316L stainless steel. *J. Nucl. Mater.* **1995**, *227*, 24–31. [CrossRef]
23. Mei, L.; Yan, D.; Xie, S.; Lei, Z.; Ge, X. Effects of Cr_2O_3 active agent on the weld process dynamic behavior and joint comprehensive properties of fiber laser welded stainless steel thick plate. *Opt. Lasers Eng.* **2020**, *128*, 106027. [CrossRef]
24. Ragavendran, M.; Vasudevan, M. Laser and hybrid laser welding of type 316L(N) austenitic stainless steel plates. *Mater. Manuf. Process.* **2020**, *35*, 922–934. [CrossRef]
25. Chen, L.; Mi, G.; Zhang, X.; Wang, C. Numerical and experimental investigation on microstructure and residual stress of multi-pass hybrid laser-arc welded 316L steel. *Mater. Des.* **2019**, *168*, 107653. [CrossRef]
26. Xue, B.; Chang, B.; Du, D. Multi-Output Monitoring of High-Speed Laser Welding State Based on Deep Learning. *Sensors* **2021**, *21*, 1626. [CrossRef] [PubMed]
27. Liang, R.; Luo, Y.; Li, Z. The effect of humping on residual stress and distortion in high-speed laser welding using coupled CFD-FEM model. *Opt. Laser Technol.* **2018**, *104*, 201–205. [CrossRef]
28. Li, Y.; Jiang, P.; Li, Y.; Mi, G.; Geng, S. Microstructure evolution and mechanical properties in the depth direction of ultra-high power laser-arc hybrid weld joint of 316L stainless steel. *Opt. Laser Technol.* **2018**, *104*, 201–205. [CrossRef]
29. Zhang, J.; Hu, K.; Zhao, J.; Duan, S.; Zhan, X. Effect of heat input on microstructure and corrosion resistance in heat affected zone of 304 stainless steel joint by laser welding. *Mater. Today Commun.* **2022**, *30*, 103054. [CrossRef]
30. Sadeh, S.; Mathews, R.; Zhang, R.; Sunny, S.; Marais, D.; Venter, A.M.; Li, W.; Malik, A. Interlayer machining effects on microstructure and residual stress in directed energy deposition of stainless steel 316L. *J. Manuf. Process.* **2023**, *94*, 69–78. [CrossRef]
31. Xu, J.; Chen, C.; Lei, T.; Wang, W.; Rong, Y. Inhomogeneous thermal-mechanical analysis of 316L butt joint in laser welding. *Opt. Laser Technol.* **2019**, *115*, 71–80. [CrossRef]
32. ISO 14175:2008 [Electronic Resource]//ISO. Available online: https://www.iso.org/ru/standard/39569.html (accessed on 20 November 2023).
33. BS EN 61326-1:2013 | 28 February 2013 | BSI Knowledge [Electronic Resource]. Available online: https://knowledge.bsigroup.com/products/electrical-equipment-for-measurement-control-and-laboratory-use-emc-requirements-general-requirements?version=standard (accessed on 20 November 2023).
34. ISO 23277:2015 [Electronic Resource]//ISO. Available online: https://www.iso.org/standard/62315.html (accessed on 20 November 2023).
35. ISO 14732:2013 [Electronic Resource]//ISO. Available online: https://www.iso.org/standard/54935.html (accessed on 20 November 2023).
36. ISO 3452-1:2021 [Electronic Resource]//ISO. Available online: https://www.iso.org/ru/standard/75696.html (accessed on 20 November 2023).
37. ISO 17636-1:2022 [Electronic Resource]//ISO. Available online: https://www.iso.org/standard/78319.html (accessed on 20 November 2023).

38. ISO 4136:2022 [Electronic Resource]//ISO. Available online: https://www.iso.org/ru/standard/81123.html (accessed on 20 November 2023).
39. ISO 5173:2023 [Electronic Resource]//ISO. Available online: https://www.iso.org/ru/standard/81724.html (accessed on 20 November 2023).
40. ISO 9016:2022 [Electronic Resource]//ISO. Available online: https://www.iso.org/ru/standard/81122.html (accessed on 20 November 2023).
41. ISO 9015-1:2001 [Electronic Resource]//ISO. Available online: https://www.iso.org/ru/standard/32821.html (accessed on 20 November 2023).
42. ISO/IEC 17025:2017 [Electronic Resource]//ISO. 2021. Available online: https://www.iso.org/ru/standard/66912.html (accessed on 20 November 2023).
43. ISO 5143:1977 en [Electronic Resource]. Available online: https://www.nen.nl/en/iso-5143-1977-en-73493 (accessed on 20 November 2023).
44. ISO 17639:2022 [Electronic Resource]//ISO. Available online: https://www.iso.org/ru/standard/81124.html (accessed on 20 November 2023).
45. ISO 3651-1:1998 [Electronic Resource]//ISO. Available online: https://www.iso.org/ru/standard/9096.html (accessed on 20 November 2023).
46. ISO 14577-1:2015 [Electronic Resource]//ISO. Available online: https://www.iso.org/ru/standard/56626.html (accessed on 20 November 2023).
47. ISO 3651-2:1998 [Electronic Resource]//ISO. Available online: https://www.iso.org/ru/standard/9098.html (accessed on 20 November 2023).
48. Voropaev, A.A.; Protsenko, V.G.; Anufriyev, D.A.; Kuznetsov, M.V.; Mukhin, A.A.; Sviridenko, M.N.; Kuryntsev, S.V. Influence of Laser Beam Wobbling Parameters on Microstructure and Properties of 316L Stainless Steel Multi Passed Repaired Parts. *Materials* **2022**, *15*, 722. [CrossRef]

materials

MDPI

Article

Effect of Preheating and Post-Heating on the Microstructures and Mechanical Properties of TC17-Ti_2AlNb Joint with Electron Beam Welding

Lihang Li *, Pengfei Fu, Bochao Lin and Xuedong Wang

Key Laboratory of Science and Technology on Power Beam Processes, AVIC Manufacturing Technology Institute, Beijing 100024, China; fupengfei97@163.com (P.F.); 18524439219@163.com (B.L.); wxue2012@yeah.net (X.W.)
* Correspondence: llh627625@126.com

Abstract: To enhance welding quality and performance, preheating and post-heating are usually employed on high-temperature materials, concurrently with welding. This is a novel technique in vacuum chamber electron beam welding (EBW). TC17 and Ti_2AlNb alloys are the hot topics in aero-engine parts, and the welding of dissimilar materials is also a broad prospect. To settle welding cracks of Ti_2AlNb, EBW with preheating and post-heating was investigated on TC17 and Ti_2AlNb dissimilar alloy, which improved the manufacturing technology on high-temperature materials. The dissimilar joint no longer had cracks after preheating, which exhibited excellent welding stability and metallurgical homogeneity, and preheating and annealing had an important effect on mechanical properties. The joint strength after 630 °C annealing is higher than that of TC17 alloy base metal (BM) and other annealing temperatures, reaching 1169 MPa at room temperature and 894 MPa at 450 °C tensile condition. The joint plasticity after 740 °C annealing is equivalent to TC17 BM. EBW with preheating improved the microstructure characteristics and enhanced the plasticity of Ti_2AlNb alloy weld and dissimilar joint, which would contribute to the application of Ti_2AlNb alloy and Ti_2AlNb dissimilar parts.

Keywords: electron beam welding (EBW); dissimilar materials; heat treatment; microstructure; mechanical property

1. Introduction

Titanium alloys are widely used in many fields, such as aerospace, chemical industry, shipbuilding and biomedical science, because of their low density; high specific strength; and comprehensive properties, such as good corrosion resistance and fatigue properties [1,2]. TC17 titanium alloy is a typical $\alpha + \beta$ two-phase titanium alloy, which is often used to manufacture key components of aircraft engines, such as shafts, compressor disks and blades, due to its high strength, good fracture toughness, high hardening depth and wide forging temperature range [3,4]. Ti_2AlNb alloy originates from Nb-modified Ti_3Al-based alloy, mainly consisting of B2, O and $\alpha 2$ phases, and it has good plasticity, fracture toughness and creep resistance, which is expected to replace nickel-based superalloy for aero-engine compressors [5–7]. In order to meet the future aero-engine demand for high-performance lightweight and high-temperature resistant materials, TC17 and Ti_2AlNb's dissimilar structure will be used as an alternative to the high-temperature part of aircraft engines. Solving the welding problem of these two alloys can not only promote the application in high thrust-to-weight ratio engines, but also be important to reduce the weight of the aircraft and improve the high-temperature service performance of the aircraft.

Due to the different chemical composition and intermetallic compound, the welding of dissimilar materials is usually prone to brittle phase, segregation and large residual stresses [8–10]. Vacuum EBW offers higher benefits due to its concentrated energy density, tiny welding deformation and high welding purity, as well as its process adaptability for welding dissimilar materials. Studies have shown that an excessive temperature gradient during Ti_2AlNb alloy welding was prone to cracking and welding defects [11]. Because of the high residual stress during the heat treatment process after welding, the secondary crack will also appear further [12,13]. The mechanical properties of the welded joints between different materials are anticipated to improve thanks to the excellent weldability of TC17 titanium alloy [14,15].

Preheating is an effective method to eliminate welding cracks for Ti_2AlNb alloy with EBW, such as multi-beam preheating and scanning preheating [16,17]. However, the multi-beam scanning preheating gradually diminishes when the weld thickness exceeds 5 mm. It is expected that the overall preheating method will perform better for larger thickness because the temperature uniformity of the material along the thickness direction is better. Furthermore, the preheating temperature needs to be matched to the post-weld annealing temperature. The preheating also has impacts on the grain and precipitation phase of the joints, which would improve the mechanical properties [17].

The effect of preheating and post-weld annealing on TC17-Ti_2AlNb dissimilar materials with EBW was investigated in this paper. The influence of various preheating and annealing temperatures on the microstructure and mechanical properties were investigated in this study. The strengthening mechanism of precipitation phase was discussed during annealing, and the preheating and annealing procedure was optimized accordingly, which helped to enhance the performance of welded joints. The better preheating effect is achieved by installing the preheating device directly in the vacuum chamber and performing electron beam welding immediately after preheating. The addition of TC17 also improves the weldability of the joints, which ultimately effectively inhibits the generation of welding cracks. This method can be further extended to the welding of any kind of refractory crack-prone materials, which is significant in engineering applications.

2. Materials and Methods

The materials of TC17 titanium alloy (Ti-5Al-4Mo-4Cr-2Zr-2Sn) and Ti_2AlNb alloy (Ti-22Al-27Nb-0.5Mo) were forged by the Western Superconducting Technologies Company and the Central Iron & Steel Research Institute from China, respectively. The plate dimension was 150 mm × 100 mm × 10 mm. The investigation of the welding process was used by ZD150-30CCV9M from China, with 30 kW power and 150 kV voltage.

The schematic of overall preheating device used in the vacuum chamber is shown in Figure 1, which was a significant part to the process for vacuum thermal protection. The device was mainly composed of a platform, heating tape and thermal insulation. The welded plate was assembled on a platform during preheating and was encircled by a heat insulation chamber to prevent heat loss. The EBW process was employed on the plate from the top after preheating. The platform was heated by heating elements below, and the thermocouples were used to monitor the temperature to guarantee the precision temperature. Before electron beam welding, the two thermocouples were to be maintained at a temperature difference of no more than 10 °C for at least 40 min. Post-weld annealing was carried out in a vacuum furnace. Different procedures were used to match the preheating and annealing temperatures. Preheating temperatures were set at 20 °C (not preheated) and 400 °C, and post-weld annealing temperatures were respectively set at 630 °C, 740 °C, and 850 °C, as shown in Table 1.

Figure 2a,b showed the typical surface characteristics of the welds. After X-ray inspection, the welding quality was conformed to meet the standard of I class, and the joints had no crack defects with preheating. The schematic diagram of the electron beam welding process is shown in Figure 2c. The surface of the test plate was polished and cleaned before welding, and then it was clamped to the welding fixture and assembled to the welding equipment platform. The optimized welding parameters were the acceleration voltage of 120 kV, welding current of 26 mA and welding speed of 10 mm/s.

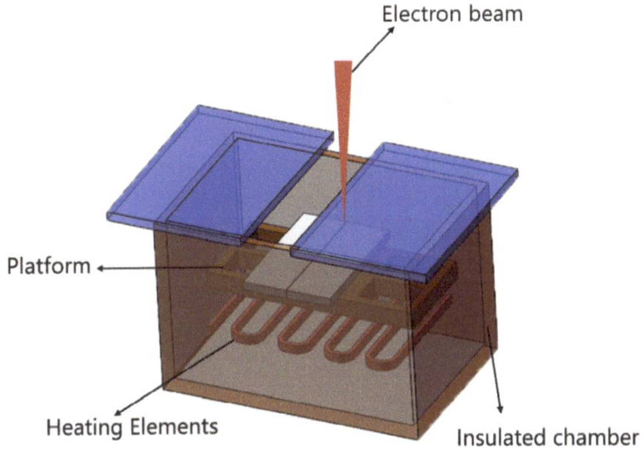

Figure 1. The schematic of preheating device in vacuum electron beam welding machine.

Table 1. The preheating and annealing process of TC17-Ti_2AlNb EBW joint. (VAC, vacuum condition; and AC, air cooling).

Number	Preheating Process	Annealing Process
NP-welded	20 °C, VAC	/
NP-630		630 °C × 2 h, AC
NP-740		740 °C × 2 h, AC
NP-850		850 °C × 2 h, AC
P-welded	400 °C, VAC	/
P-630		630 °C × 2 h, AC
P-740		740 °C × 2 h, AC
P-850		850 °C × 2 h, AC

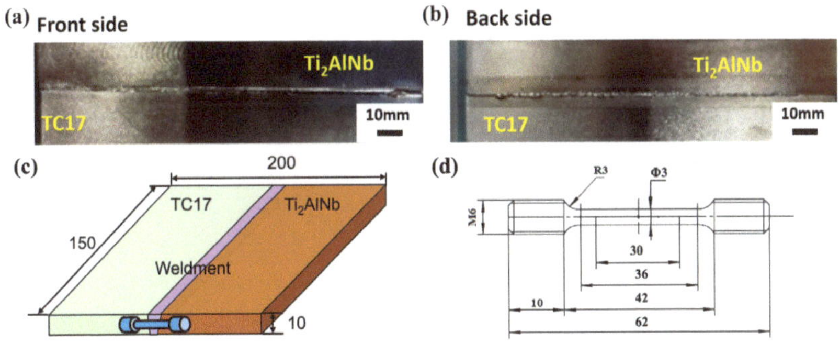

Figure 2. (**a**) Front side and (**b**) bake side of TC17-Ti_2AlNb dissimilar welds. (**c**) Schematic diagram of electron beam welding process. (**d**) Size of tensile test specimen at room temperature and 450 °C.

The specimens were prepared for metallographic observation, were ground and polished, and then were corroded using newly formulated Kohler's reagent. ZEISS field emission scanning electron microscopy (SEM) from Germany was used for high-magnification microstructure observation. The instrument used for electron backscatter diffraction (EBSD) was model JSM-7900F from Japan, and the probe was EDAX HIkari XP. The tensile tests were conducted on the CSS-2220 electronic universal testing machine from China under room temperature and high-temperature conditions. As the main service temperature of TC17 was 450 °C, the high-temperature tensile properties were chosen to be tested at 450 °C. The dimensions of the tensile specimen are shown in Figure 2d. The elongation is the percentage of the ratio of the residual elongation according to the mark ($L-L_0$) after fracture to the original mark (L_0). For the determination of elongation after fracture, the broken parts of the specimen are carefully mated together so that their axes are in the same straight line, and measures are taken to ensure that the broken parts of the specimen are in proper contact with each other before measuring the specimen's distance after fracture (L).

The finite element analysis of the welding temperature field was completed by the SYSWELD software from ESI company, France. The heat source of Gaussian three-dimensional volume was employed in modeling the welding process of TC17-Ti_2AlNb dissimilar materials, including the conventional welding and welding with preheating. The Volume of Fluid (VOF) method was used to calculate the welding velocity field. The three-dimensional model was established according to the plate dimension. The initial and boundary conditions were based on the following assumptions: (1) The material was continuous, homogeneous and isotropic. (2) The physical characteristics of the material depended on the temperature. (3) The heat dissipation under vacuum conditions was heat conduction between the test plates, and heat radiation only generated on the surface. There was no convective heat transfer. (4) The vaporization and flow of liquid metal during the experiment were not considered.

3. Results and Discussion

The morphology of the TC17-Ti_2AlNb alloy with EBW is shown in Figure 3a,b. The weld and base metal (BM) exhibited discontinuous microstructure characteristics. The welded joint comprised a BM zone near the TC17 side (TC17-BM), HAZ near the TC17 side (TC17-HAZ), fusion zone (FZ), HAZ near the Ti_2AlNb side (Ti_2AlNb-HAZ) and BM zone near the Ti_2AlNb side (Ti_2AlNb-BM), as seen in Figure 3c–e, which show the weld cracks appearing in the heat-affected zone on the Ti_2AlNb side after annealing without preheating. The locations where cracks were found are marked with a yellow box in Figure 3a. This crack appeared during the annealing process, sprouting from the Ti_2AlNb-HAZ and expanding along the intergranularity to both sides. The direction of the crack extension is basically perpendicular to the weld depth direction, indicating that the crack is generated due to the large temperature gradient between the FZ and Ti_2AlNb-BM.

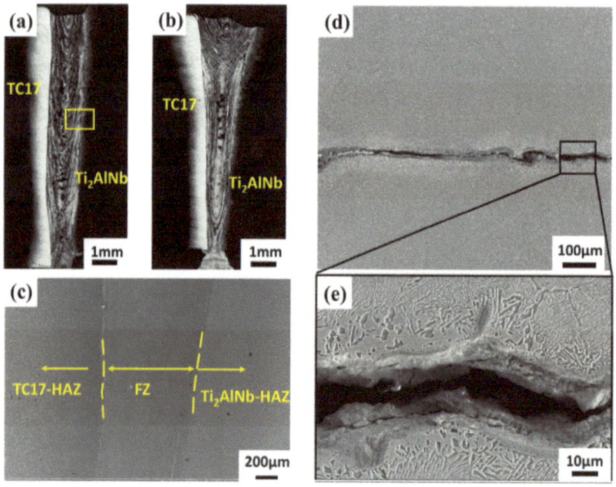

Figure 3. Cross-section morphology of TC17-Ti_2AlNb welded joint: (**a**) macro view of non-preheated joint and (**b**) preheated joint by optical microscope. (**c**) Micro view by SEM. The cracks of TC17-Ti_2AlNb welded joint on Ti_2AlNb side after annealing without being preheated by SEM: (**d**) macro view and (**e**) micro view.

The microstructure of the TC17 alloy shows an equiaxial α phase distributed on the β phase (Figure 4a). The microstructure of the Ti_2AlNb alloy is characterized by an α phase and O phase distributed on the β-phase base (Figure 4b). The dissimilar joint microstructures of EBW (as-welded) and EBW with heat treatment are shown in Figures 4 and 5. The microstructure's evolution is mainly attributed to the rapid melting and solidification process during EBW, but the preheating prior to welding had little impact on the microstructure. The weld (Figure 4c) retained a significant amount of β phase due to the fast cooling after welding. Figure 4d shows the microstructure of the FZ after annealing. O or α phases precipitated from the β phase were uniformly distributed in the weld, with the majority of the O phase precipitating from the α phase's grain boundaries and creating a marginal O phase around the α phase [18,19].

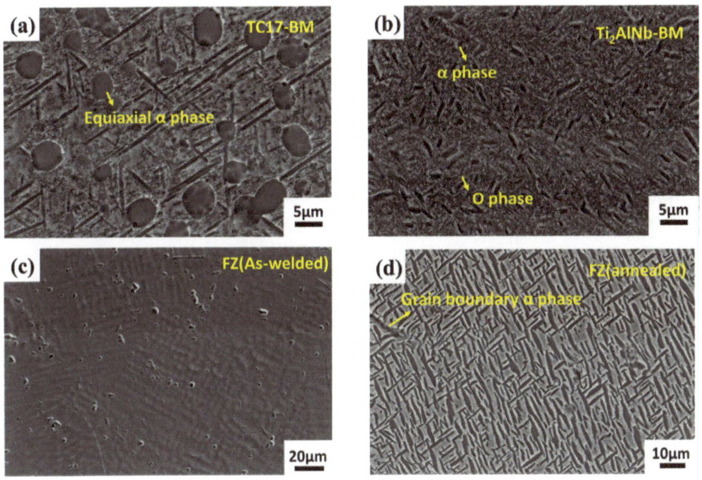

Figure 4. Microstructure of base metal by SEM: (**a**) TC17 alloy and (**b**) Ti_2AlNb alloy. Microstructure of TC17-Ti_2AlNb electron beam-welded joints for FZ by SEM: (**c**) as-welded and (**d**) annealed.

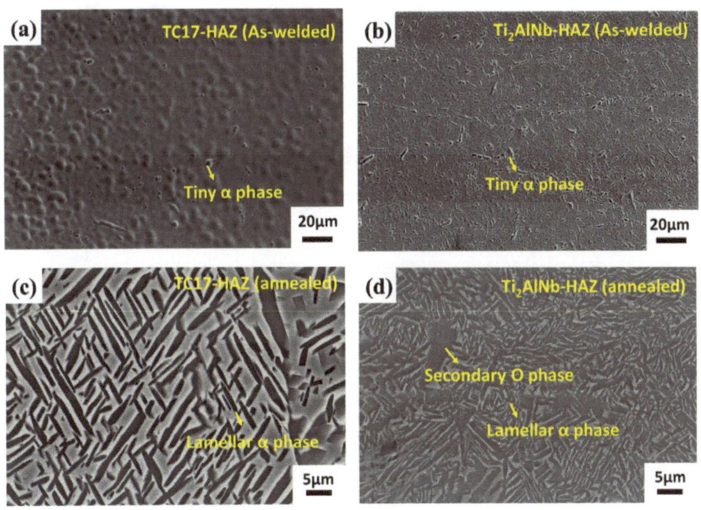

Figure 5. Microstructure of TC17-Ti_2AlNb electron beam-welded joints for HAZ by SEM: (**a**) TC17-HAZ (as-welded), (**b**) Ti_2AlNb-HAZ (As-welded), (**c**) TC17-HAZ (annealed) and (**d**) Ti_2AlNb-HAZ (annealed).

A small amount of α phase existed in HAZ of the dissimilar joint (Figure 5a,b). The microstructure displayed a more pronounced variability when the precipitation phase gradually appeared after annealing. The microstructures of TC17-HAZ (Figure 5c) were a lot of precipitated strips of the α phase. The reticulated B2 + α2 + O microstructures existed in Ti_2AlNb-HAZ (Figure 5d) with a high proportion of O phase.

In contrast to conventional welding, preheating increases the temperature and improves the homogeneity of the joints' temperature field, which had a significant impact on the welding thermal cycling. The HAZ exhibited a longer post-weld cooling time and a slower cooling rate, which was verified from the modeling in Figure 6. Figure 6a,b, respectively, show the simulation results of the temperature field under two conditions: non-preheated and preheated. From Figure 6c, the peak temperature of the location with 3 mm distance from the welding center increased at least 200 °C in the preheating area during welding. The preheated welding took 19.5 s as opposed to 4.2 s for the conventional welding under the same conditions, cooling to 600 °C. To better simulate the changes in keyhole and melting pool during the EBW process, the Volume of Fluid (VOF) method was used to calculate the welding velocity field [20]. The model was optimized and validated by comparing the shapes in the model with the cross-section of the welded joint. It could be found from Figure 6d,e and Table 2 that the shapes of the transverse cross-sections of the weld obtained were in accordance with the experimental results for both conditions, and the calculation errors were restricted to 7.8%.

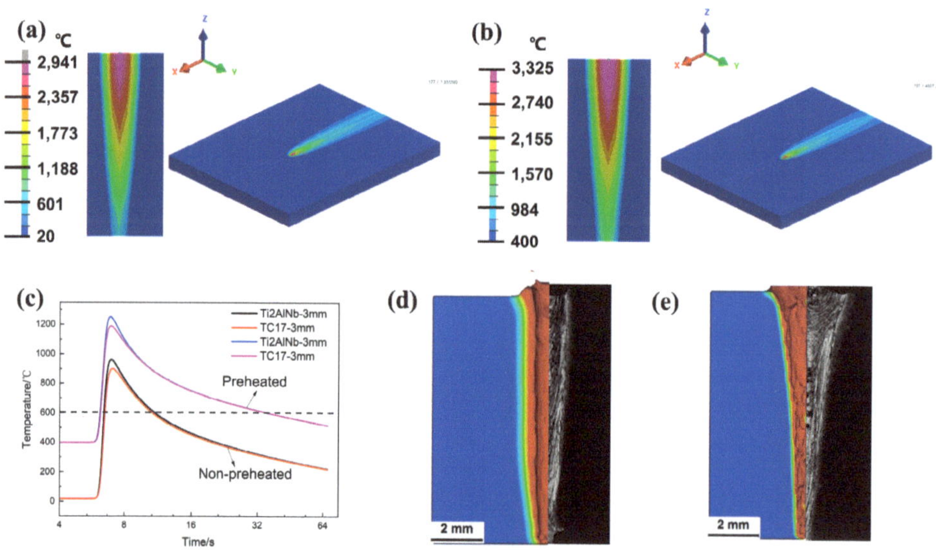

Figure 6. The temperature field of (**a**) non-preheated and (**b**) preheated. (**c**) Temperature field variation at 3 mm nodes from the center of the weldment on both sides after welding. The calculational and experimental results of the transverse cross-sections of the weld: (**d**) non-preheated and (**e**) preheated.

Table 2. The comparison of the calculational and experimental results of the weld dimensions.

Location	Non-Preheated (Figure 6d)			Preheated (Figure 6e)		
	Experimental Result/mm	Calculational Result/mm	Error	Experimental Result/mm	Calculational Result/mm	Error
Width of the top of the weld	1.26	1.21	4.3%	3.23	2.99	7.8%
Width of the intermediate part of the weld	0.98	1.03	5%	1.26	1.25	1.0%
Width of the bottom of the weld	0.59	0.62	4.3%	0.62	0.59	4.3%

The results of the grain size characterized by EBSD also supported the results of the simulation (Figure 7). The grain sizes in each region increased significantly after welding compared to the base metal. The average grain sizes of TC17-BM and Ti_2AlNb-BM are about 15 μm and 2 μm, respectively (Figure 7a). In general, the grain size basically shows a decreasing tendency with the increase in the annealing temperature, and the change is especially obvious in the Ti_2AlNb-HAZ. However, under the condition of non-preheated, the FZ and TC17-HAZ of the welded joints were characterized by small fluctuating changes.

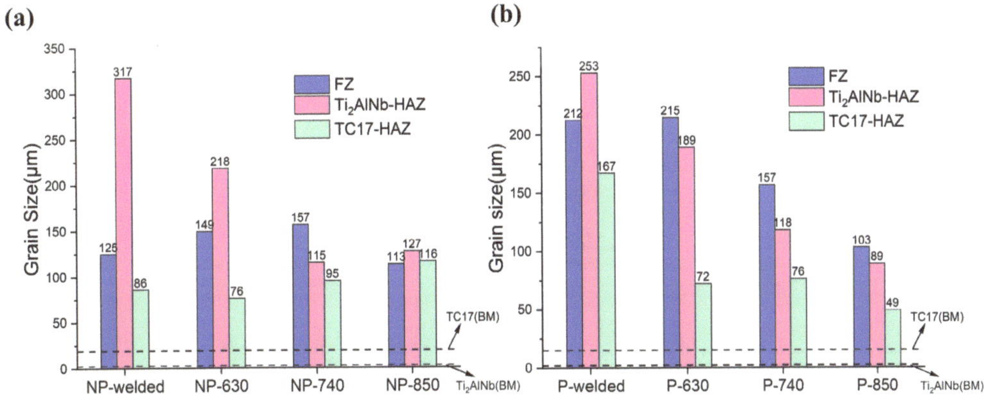

Figure 7. Grain size distribution of TC17-Ti_2AlNb dissimilar joints: (**a**) non-preheated and (**b**) preheated.

Comparing the grain size under NP-welded and P-welded conditions, the grains were more uniform after preheating (Figure 7b). Preheating lowered the gradient of temperature throughout the welding zone, which slowed post-weld cooling and lengthened the time for grain nucleation and growth. The grains for the P-welded conditions were coarser than those of the NP-welded conditions, which were more uniform in the HAZ and FZ. As the post-weld annealing process moved on, the region's average grain size tended to decrease while simultaneously becoming more uniform.

The tensile properties of TC17-Ti_2AlNb joints at room temperature are shown in Figure 8a,b. The tensile strength was attained for 1169 MPa under the P-630 condition, which was roughly equivalent to the strength of the Ti_2AlNb base metal. The preheated joint's tensile strength was slightly higher at 850 °C than at 740 °C, but the joint's elongation displayed the opposite tendency. The preheating method effectively improved the tensile strength of the joints at room temperature by 5%, 5.2% and 3.3% at 630 °C, 740 °C and 850 °C, respectively. The tensile fracture at 740 °C annealing was in the TC17-HAZ, while another two temperatures' fractures were both in the FZ, as shown in Figure 9a,b.

Figure 8c,d demonstrate the tensile properties at 450 °C of the TC17-Ti_2AlNb joints. It could be found that the strength of TC17-Ti_2AlNb joints at 450 °C varies with different annealing temperatures in the same way as the room temperature, showing a slightly higher strength at 630 °C and the lowest strength at 740 °C. The tensile strength reached 894 MPa under the P-630 condition. Compared with non-preheated process, the tensile strength of the dissimilar joints increased by 4%, 5.3% and 4% at 630 °C, 740 °C and 850 °C, respectively. The elongation reached over 18% at 740 °C and 850 °C, which is comparable to the Ti_2AlNb base metal. The elongation is 15.8% for NP-630 and 15.5% for P-630. The fractures at 450 °C are all located in the TC17-BM, as shown in Figure 9c. The preheating proved to improve the high-temperature mechanical properties of the dissimilar joints effectively.

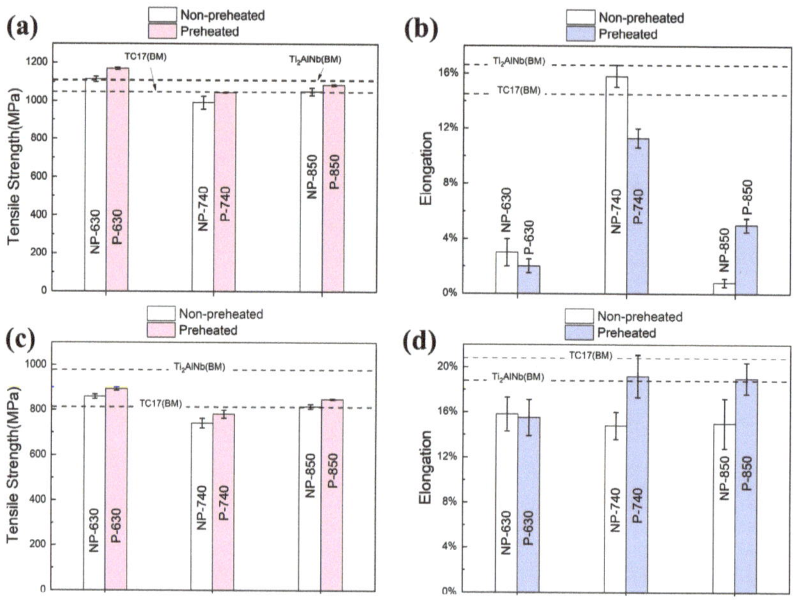

Figure 8. Tensile properties of TC17-Ti_2AlNb joints at room temperature: (**a**) tensile strength. (**b**) Elongation and high temperature tensile properties of TC17-Ti_2AlNb joints at 450 °C: (**c**) tensile strength and (**d**) elongation.

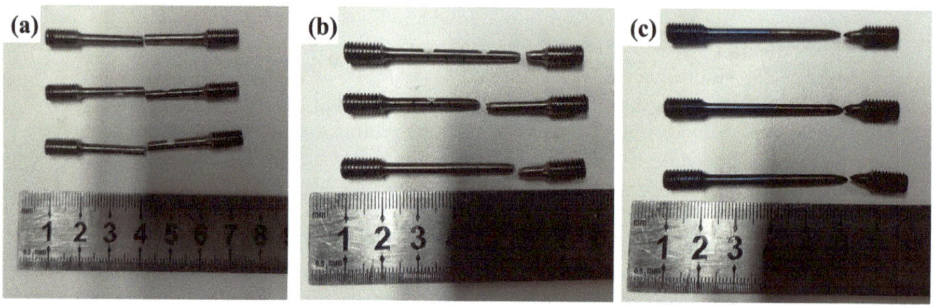

Figure 9. Tensile specimen of TC17-Ti_2AlNb joints: (**a**) room temperature for 630 °C × 2 h AC and 850 °C × 2 h AC, (**b**) room temperature for 740 °C × 2 h AC and (**c**) high temperature at 450 °C.

The increased mechanical properties of the dissimilar joints were attributed to the improvement of microstructure and grain structure. Figure 10a shows, for the condition of the 850 °C treatment, the proportion of phase composition in each area of non-preheated welding and preheated welding characterized by EBSD. The most noticeable change was that the preheating reduced the proportion of the α phase in the FZ from 45.4% to 16.6%. Figure 10b–d show the phase ratio of the TC17-Ti_2AlNb joint for three annealing temperatures, respectively. Preheated and non-preheated conditions are both presented as a large proportion of the β phase after welding. However, the ratio of O phase that precipitated from the β-phase matrix would differ after various annealing temperatures, which would affect the mechanical properties of the dissimilar joints.

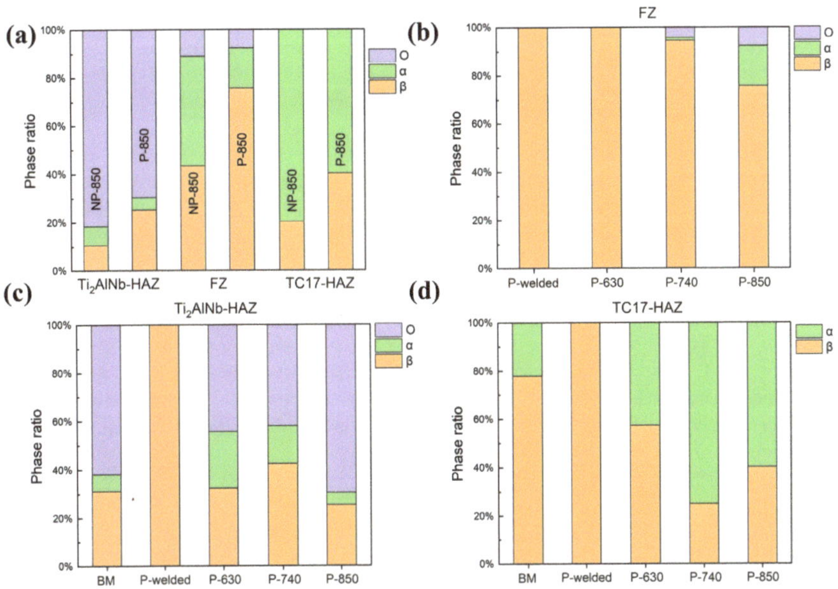

Figure 10. Histograms of phase ratios in TC17-Ti_2AlNb joints in different conditions. (**a**) Comparison of preheated and non-preheated in different regions of the weld. Comparison of different conditions for (**b**) FZ, (**c**) Ti_2AlNb-HAZ and (**d**) TC17-HAZ.

The tensile properties of dissimilar joints with different annealing temperatures are closely related to the microstructures and phase ratios. The FZ maintained β-phase characteristics after 630 °C annealing, showing higher strength but lower plasticity at room temperature, and the fracture location was also in the FZ zone. The precipitated phases first appeared in FZ after 740 °C annealing. Both the α phases and O phases were tiny, and the β phase remained dominant in the zone. However, after 850 °C annealing, the O and α phases in the FZ zone reached 7.7% and 16.6%, respectively. Generally, Ti_2AlNb-based alloys with single O phases had lower strength and plasticity due to the large proportion of O/O grain boundaries [21]. The α/B2 grain boundary and α/O grain boundary, however, were more easily to develop cracks when the α phase emerged as a brittle phase in FZ, which ultimately caused the joint to fracture with the crack growth [22,23].

The HAZ on both sides grew a large proportion of precipitated phases during annealing. The precipitation of α phase in the TC17-HAZ zone showed a trend of increasing and then decreasing, with the highest proportion of α phase after 740 °C annealing and reaching 75%. This resulted in the welded joints' rupturing on the TC17 side, with a low room temperature tensile strength but high elongation. The reason could be attributed to the sub-stable β phase decomposing and secondary needle-like α phase precipitating after 630 °C and 740 °C annealing [23,24]. However, when the annealing temperature rose to 850 °C, which was located at the higher temperature of the α + β phase region, the annealing process resulted in an α→β phase transformation, converting the majority of the α phase into the β phase.

In Ti_2AlNb-HAZ, the annealing temperature of 630~850 °C was also in the B2 + O two-phase region from the Ti_2AlNb phase diagram [25]. The driving force of O-phase nucleation and growth in the joint increased with the temperature enhancing, and the resulting O phase would be spread around the α2 phase, inhibiting the α2 phase growth further [23,26]. The superior performance of Ti_2AlNb under a high temperature resulted in a welded joint with a more excellent high-temperature strength than the TC17 base metal, eventually causing the fractures in TC17-BM [27]. The increase in the proportion of the α phase in the TC17 alloy is the main reason for the increase in plasticity. Moreover, 630 °C just reaches the α + β phase transforming temperature of the TC17 alloy, at which less α phase is generated after heat treatment. The effect of preheating on the α phase is less pronounced compared to higher temperatures (740 °C and 850 °C), so that the high-temperature plasticity remains essentially unchanged.

Figure 11a,b show the tensile fracture at room temperature when the annealing temperature is 630 °C and 850 °C. The morphology showed typical microporous intensive fracture characteristics, and the fracture was located in FZ. Lacerate ribs consisting of dimples can be observed at the fracture, indicating that the welded joints have a certain plastic deformation capacity at room temperature, and, eventually, a ductile fracture was caused by local plastic deformation. The tensile fracture location at 740 °C is located on the TC17 side, and an obvious necking phenomenon occurs in Figure 11c,d. The proportion of the α phase on the TC17 side after annealing at this temperature increased significantly, reaching 74% (Figure 10d), so that the deformation in the tensile tests was more distributed to the hexagonal structure, which was more favorable to the opening of multiple slip systems, and eventually manifested as a significant increase in the plasticity of the joint. However, the strength of the α phase is not as strong as that of the β phase, so the tensile strength of the joint is significantly reduced at 740 °C, as seen in Figure 8a. The high-temperature tensile fractures in Figure 11e,f were all located on the TC17 side, and the morphology was similar to that of Figure 11a,b. The reason for this phenomenon was mainly due to the precipitation of secondary α phase during the high-temperature test so that the α phase dominated the microstructure of the TC17 side.

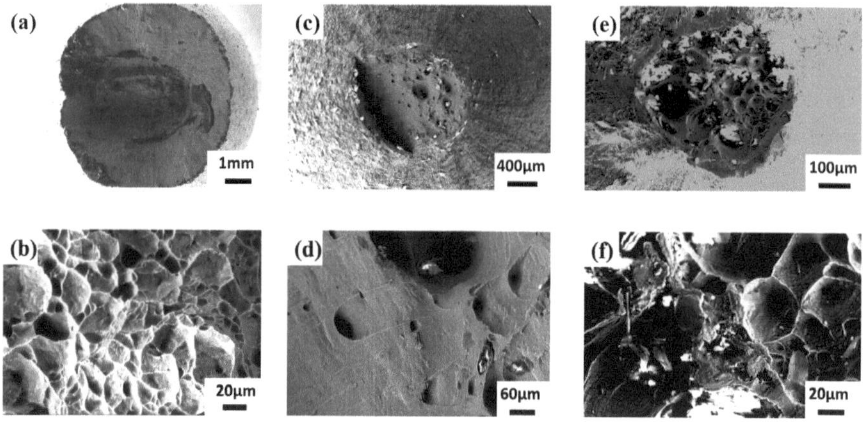

Figure 11. Tensile fracture of TC17-Ti_2AlNb joints by SEM: (**a**,**b**) room temperature for 630 °C × 2 h AC and 850 °C × 2 h AC, (**c**,**d**) room temperature for 740 °C × 2 h AC and (**e**,**f**) high temperature at 450 °C.

Due to the longer post-weld cooling time caused by preheating, the proportion of annealing precipitation phases in the joints was significantly reduced. Figure 12 showed the grain orientation polar chart of TC17-Ti_2AlNb joints with non-preheated and preheated welding processes. The grain orientation distribution of the joint without preheating was more stochastic, indicating that the growth of grains was not obvious in terms of directionality. The greatest texture strength was 15.622 for preheated welding, with grain growth being more concentrated in the (001) direction. Compared with the non-preheated welding, the significant increase in texture strength indicated that the welding microstructure exhibited significant directionality under preheated welding. The change in grain orientation could also be attributed to the grain remaining in the high temperature zone for a long time. As a result, the cooling rate was slower, and the grain was more likely to be selective in growth.

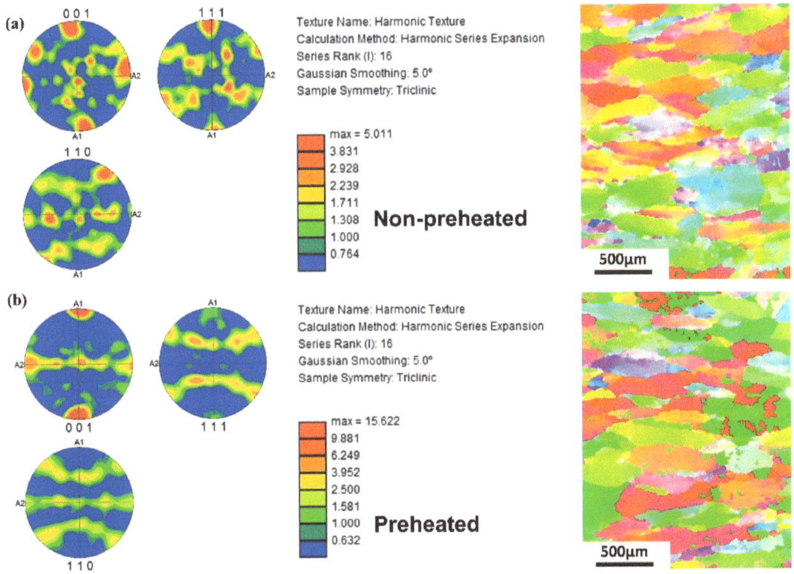

Figure 12. EBSD orientation polar map and grain structure of TC17-Ti_2AlNb welded joints for (**a**) non-preheated and (**b**) preheated.

In general, the microstructure of FZ usually emerged as the atomically disordered β phase after solidification, and it was then further ordered by a solid-state phase transformation during subsequent cooling [17,28]. In contrast to preheated conditions, the rapid cooling rate without preheating prevented the orderly transformation of β phase, leaving a more chaotic sub-stable β phase. The O and α phases precipitated after annealing mainly originated from the decomposition of sub-stable β phase. The transforming of B2→O for Ti_2AlNb side and β→α for TC17 side would arise in FZ when the annealing temperature reached 600 °C. Therefore, the slow cooling rate and less residual sub-stable β phase of EBW with preheating would lead to a decrease in the proportion of O phase and α phase after annealing. The reduction of the α phase was the main reason for the increase in the joint's strength under preheating conditions.

The tensile strength and high-temperature plasticity of the TC17-Ti_2AlNb joint could be effectively improved by appropriate combination of the preheating and post-weld annealing. As the TC17 titanium alloy with greater weldability was added, we could raise the joint's tensile strength and improve its mechanical properties in comparison to Ti_2AlNb electron beam-welded joints individually [19].

4. Conclusions

(1) EBW with preheating was employed TC17 and Ti_2AlNb dissimilar alloy, which showed excellent welding metallurgical compatibility with non-defects joint after preheating. The weld has a mainly β-phase microstructure, and a few of O phases and α phases were generated in the weld only after annealing. But a lot of O and α phases precipitated in HAZ on TC17 and Ti_2AlNb sides.

(2) The tensile strengths of dissimilar joints under room temperature and 450 °C were equivalent to those of TC17 BM, and the strengths after 740 °C annealing are lower than those after 630 °C and 850 °C annealing. The tensile fracture after 740 °C annealing was in TC17 HAZ due to the low strength, which reaches 75% proportion of α phase in HAZ. The dissimilar joint achieved the highest strength after 630 °C annealing, reaching 1169 MPa at room temperature and 894 MPa at 450 °C tensile condition.

(3) The preheating increased the peak temperature in HAZ and prolonged the welding cooling period from the numerical simulation, which decreased almost 30% of brittle α phase, contributing to an increase of 5% in tensile strength. Preheating would effectively increase the strength and plasticity of the dissimilar joint.

Author Contributions: Conceptualization, L.L. and P.F.; methodology, L.L.; software, X.W.; validation, L.L., P.F. and B.L.; formal analysis, L.L. and P.F.; investigation, L.L.; resources, L.L.; data curation, L.L.; writing—original draft preparation, L.L.; writing—review and editing, L.L. and P.F.; visualization, L.L.; supervision, P.F., B.L. and X.W.; project administration, L.L.; funding acquisition, P.F. All authors have read and agreed to the published version of the manuscript.

Funding: The authors would like to thank the financial support of National Science and Technology Major Project (J2022-VII-0003-0045).

Institutional Review Board Statement: Not applicable.

Informed Consent Statement: Not applicable.

Data Availability Statement: Data are contained within the article.

Acknowledgments: The authors sincerely express their thanks to the team members of this project, Zhenyun Tang, Zhiyong Mao and An Zhang. The authors also appreciate Jiancheng Xu for measuring the mechanical properties.

Conflicts of Interest: The authors declare no conflict of interest.

References

1. Auwal, S.T.; Ramesh, S.; Yusof, F.; Manladan, S.M. A review on laser beam welding of titanium alloys. *Int. J. Adv. Manuf. Technol.* **2018**, *97*, 1071–1098. [CrossRef]
2. Zhao, Q.Y.; Sun, Q.Y.; Xin, S.W.; Chen, Y.N.; Wu, C.; Wang, H.; Xu, J.W.; Wan, M.P.; Zeng, W.D.; Zhao, Y.Q. High-strength titanium alloys for aerospace engineering applications: A review on melting-forging process. *Mater. Sci. Eng. A—Struct. Mater. Prop. Microstruct. Process.* **2022**, *845*, 143260. [CrossRef]
3. García, J.M.; Gaslain, F.; Morgeneyer, T.F. On the effect of a thermal treatment on the tensile and fatigue properties of weak zones of similar Ti17 linear friction welded joints and parent material. *Mater. Charact.* **2020**, *169*, 110570. [CrossRef]
4. Ben Boubaker, H.; Mareau, C.; Ayed, Y.; Germain, G.; Tidu, A. Impact of the initial microstructure and the loading conditions on the deformation behavior of the Ti17 titanium alloy. *J. Mater. Sci.* **2020**, *55*, 1765–1778. [CrossRef]
5. Goyal, K.; Sardana, N. Mechanical Properties of the Ti_2AlNb Intermetallic: A Review. *Trans. Indian Inst. Met.* **2021**, *74*, 1839–1853. [CrossRef]
6. Malecka, J.; Rozumek, D. Metallographic and Mechanical Research of the O-Ti_2AlNb Alloy. *Materials* **2020**, *13*, 3006. [CrossRef]
7. Yamaguchi, M.; Inui, H.; Ito, K. High-temperature structural intermetallics. *Acta Mater.* **2000**, *48*, 307–322. [CrossRef]
8. Anawa, E.M.; Olabi, A.G. Control of welding residual stress for dissimilar laser welded materials. *J. Mater. Process. Technol.* **2008**, *204*, 22–33. [CrossRef]
9. Kim, H.T.; Kil, S.C. Research Trend of Dissimilar Metal Welding Technology. In *Computer Applications for Modeling, Simulation, and Automobile*; Springer: Berlin/Heidelberg, Germany, 2012; Volume 341, pp. 199–204.
10. Galeev, R.M.; Valiakhmetov, O.R.; Safiullin, R.V.; Imaev, V.M.; Imaev, R.M. Microstructure and Properties of a Titanium Alloy-Orthorhombic Titanium Aluminide Layered Composite. *Phys. Met. Metallogr.* **2009**, *107*, 312–316. [CrossRef]
11. Zhang, K.; Ni, L.; Lei, Z.; Chen, Y.; Hu, X. In situ investigation of the tensile deformation of laser welded Ti_2AlNb joints. *Mater. Charact.* **2017**, *123*, 51–57. [CrossRef]
12. Cai, D.; Chen, J.; Mao, X.; Hao, C. Reheat cracking in Ti_2AlNb alloy resistance spot weldments. *Intermetallics* **2013**, *38*, 63–69. [CrossRef]
13. Leo, P.; D'Ostuni, S.; Nobile, R.; Mele, C.; Tarantino, A.; Casalino, G. Analysis of the Process Parameters, Post-Weld Heat Treatment and Peening Effects on Microstructure and Mechanical Performance of Ti-Al Dissimilar Laser Weldings. *Metals* **2021**, *11*, 1257. [CrossRef]
14. Hsieh, C.T.; Shiue, R.K.; Huang, R.T.; Tsay, L.W. The effect of post-weld heat treatment on the microstructure and notched tensile fracture of Ti-15V-3Cr-3Al-3Sn to Ti-6Al-4V dissimilar laser welds. *Mater. Sci. Eng. A—Struct. Mater. Prop. Microstruct. Process.* **2016**, *653*, 139–146. [CrossRef]
15. Boyat, X.; Ballat-Durand, D.; Marteau, J.; Bouvier, S.; Favergeon, J.; Orekhov, A.; Schryvers, D. Interfacial characteristics and cohesion mechanisms of linear friction welded dissimilar titanium alloys: Ti-5Al-2Sn-2Zr-4Mo-4Cr (Ti17) and Ti-6Al-2Sn-4Zr-2Mo (Ti6242). *Mater. Charact.* **2019**, *158*, 109942. [CrossRef]
16. Li, L.H.; Fu, P.F.; Zhao, T.; Tang, Z.Y.; Mao, Z.Y. Effect of Preheating on the Microstructure Evolution and Mechanical Properties of Electron Beam Welded Ti_2AlNb Alloy. *J. Mater. Eng. Perform.* **2023**, *32*, 3648–3657. [CrossRef]

17. Panov, D.; Naumov, S.; Stepanov, N.; Sokolovsky, V.; Volokitina, E.; Kashaev, N.; Ventzke, V.; Dinse, R.; Riekehr, S.; Povolyaeva, E.; et al. Effect of pre-heating and post-weld heat treatment on structure and mechanical properties of laser beam-welded Ti_2AlNb-based joints. *Intermetallics* **2022**, *143*, 107466. [CrossRef]
18. Muraleedharan, K.; Nandy, T.K.; Banerjee, D.; Lele, S. Phase stability and ordering behaviour of the O phase in Ti-Al-Nb alloys. *Intermetallics* **1995**, *3*, 187–199. [CrossRef]
19. Chen, W.; Chen, Z.Y.; Wu, C.C.; Li, J.W.; Tang, Z.Y.; Wang, Q.J. The effect of annealing on microstructure and tensile properties of Ti–22Al–25Nb electron beam weld joint. *Intermetallics* **2016**, *75*, 8–14. [CrossRef]
20. Wang, X.; Fu, P.; Qingyu, S. Mathematical Delineation of the Thermophysical Process of Electron Beam Welding. *Rare Met. Mater. Eng.* **2011**, *40*, 134–137.
21. Boehlert, C.J. Part III. The tensile behavior of Ti-Al-Nb O + Bcc orthorhombic alloys. *Metall. Mater. Trans. A* **2001**, *32*, 1977–1988. [CrossRef]
22. Gogia, A.K.; Baligidad, R.G.; Banerjee, D. Toughening and creep in multiphase intermetallics through microstructural control. *Sadhana* **2003**, *28*, 677–693. [CrossRef]
23. Goyal, K.; Sardana, N. Phase stability and microstructural evolution of Ti_2AlNb alloys—A review. In *Materials Today: Proceedings*; Elsevier: Amsterdam, The Netherlands, 2021; Volume 41, pp. 951–968.
24. Yu, B.B.; Chen, Z.Y.; Zhao, Z.B.; Liu, J.R.; Wang, Q.J.; Li, J.W. Microstructure and mechanical properties of electron beam weldment of titanium alloy Tc17. *Acta Metall. Sin.* **2016**, *52*, 831–841.
25. Boehlert, C.J.; Majumdar, B.S.; Seetharaman, V.; Miracle, D.B. Part I. The microstructural evolution in Ti-Al-Nb O + Bcc orthorhombic alloys. *Metall. Mater. Trans. A—Phys. Metall. Mater. Sci.* **1999**, *30*, 2305–2323. [CrossRef]
26. Chen, X.; Xie, F.Q.; Ma, T.J.; Li, W.Y.; Wu, X.Q. Microstructural evolution and mechanical properties of linear friction welded Ti_2AlNb joint during solution and aging treatment. *Mater. Sci. Eng. A* **2016**, *668*, 125–136. [CrossRef]
27. Yamabe-Mitarai, Y.; Kuroda, S.; Motohashi, N.; Matsumoto, H.; Miyamoto, G.; Chandiran, E.; Yoshida, Y.; Itsumi, Y. Effect of Forging Temperature on Microstructure Evolution and Tensile Properties of Ti-17 Alloys. *Mater. Trans.* **2019**, *60*, 1733–1739. [CrossRef]
28. García, J.M.; Esin, V.A.; Morgeneyer, T.F. Strength, fatigue strength and toughness of dissimilar Ti17-Ti64 linear friction welded joints: Effect of soft surface contamination and depletion of α precipitates. *Mater. Sci. Eng. A—Struct. Mater. Prop. Microstruct. Process.* **2021**, *799*, 139989. [CrossRef]

materials

MDPI

Article

Multiphysics Study of Thermal Profiles and Residual Stress in Welding

Yousung Han

Department of Mechatronics Engineering, Incheon National University, 119 Academy-ro, Yeonsu-gu, Incheon 22012, Republic of Korea; yshan@inu.ac.kr

Abstract: One of the effects of welding is residual stress. Welding involves complex tests concerning differences in values of the mechanical parameters of its regions as an effect of residual stress. Such multiphysics characteristics of welding pose a challenge in predicting residual stress. In the present study, a thermo-mechanical constitutive model considering phase transformation and transformation plasticity is implemented in the numerical model in ABAQUS user subroutines. In order to consider phase evolution in welding, the metallurgical parameters for Leblond's phase equation were obtained from the calibration of DH36 steel with a CCT diagram. In addition, the effects of welding speed on thermal profiles and residual stress generation were investigated. Analysis has suggested that the width of the heat-affected zone (HAZ) decreases with an increase in welding speed, and the phase fraction is significantly affected by this kind of parameter. Such phase transformation has led to the generation of a compressive stress in the fusion zone (FZ) and HAZ. The volume difference between coexisting phases produces a compressive stress in cooling, and its magnitude was increased with martensite increasing.

Keywords: residual stress; finite element analysis; phase transformation; transformation plasticity

1. Introduction

Welding involves complex material behaviors with respect to the interaction of mechanical, thermal, and metallurgical processes. The phase volume fraction of steel is determined by thermal profiles. Conversely, phase transformation leads to deformation due to volume dilation among the phases. In addition, during the welding process, the latent heat and heat capacity of a material influence temperature. Such multiphysical phenomena should be considered for more precise prediction of residual stress in a welded structure.

Iron atoms in steel are arranged in either a body-centered cubic (BCC) or face-centered cubic (FCC) structure, depending on temperature. The atomistic arrangement of ferritic steels (i.e., ferrite, bainite, or martensite) is a BCC structure at room temperature. However, when temperature exceeds the eutectoid temperature, steel has FCC structures and becomes austenite. During the heating process in welding, above the eutectoid temperature, the atomistic arrangement of the iron atoms is switched from BCC to FCC structure. Conversely, in the cooling process, austenitic steel transforms into ferrite, bainite, or martensite, depending on its cooling rate. In such phase transformations, plastic flows are generated even at the stress lower than the yield stress. This is known as transformation plasticity.

Many studies have been performed to understand the effects of phase transformation on microstructure evolution and material properties [1–9]. For example, Tsirkas et al. [10] developed the numerical model for laser welding simulations considering phase transformation using the Johnson–Mehl–Avrami law and the Koistinen–Marburger law. Rong et al. [1] examined the solid phase transformation in laser welding of EH36 steel. The effects of solid phase transformation on material properties, deformation, and residual stress were investigated. Lee and Chang [2] performed sequentially coupled thermomechanical simulations to study welding-induced residual stress in consideration of phase transformation.

Most of these papers captured only phase transformation, and transformation plasticity was not explicitly considered. Leblond et al. [11,12] proposed a thermo-mechanical model incorporated with transformation plasticity. This constitutive model could predict thermal-induced residual stress more precisely.

A range of studies have been conducted to understand the influence of welding process parameters on mechanical properties, and thus the structural reliability of welded parts [13–20]. Ferro et al. [21] conducted FE simulations to examine the influence of different heat source models on thermal profiles in welding. Gannon et al. [22] performed FE simulations to examine how the welding sequence affects residual stress and warpage. Huang et al. [23] performed FE simulations to investigate the influence of the welding sequence on thermal-induced distortion, and proposed the optimal welding method to reduce buckling in the fabrication of thin plate. Biswas et al. [24] performed FE simulations to study the effects of the welding sequence on warpage and its magnitude in the welding of stiffened panels. Their results show that the welding sequence has great influence on the distortion of stiffened panel.

The influence of the heat source on residual stress was studied in several approaches [20,25–33]. Huang et al. [15] performed X-ray diffraction (XRD) analysis on welded structures to investigate the effects of strain hardening and annealing on the residual stress distribution of thin plates. Kong et al. [34] examined the suitability of double ellipsoidal and cylindrical heat source models for the numerical study of hybrid laser GMA welding. Benyounis et al. [35] conducted experimental studies to investigate the influence of welding parameters such as power, deposition speed, and focal position on the structural reliability of a welded structure to find the optimal welding conditions. The effects of welding speed on microstructural features and corrosion behavior were also investigated for the welding of AISI 201 stainless steel [36]. Microscopic analyses showed that the amount of grain coarsening decreases with an increase in welding speed. Zhang et al. [37] performed experimental analysis on the laser welding of stainless steel plate to investigate the effects of focal position and welding deposition speed on the bead geometry. The influence of welding speed on microstructural features were also analyzed for the welding of dissimilar aluminum and copper joints [38]. Analyses expressed that extremely high welding speed could suppress the formation of brittle compounds which produces a strong weld joint.

An implementation of thermomechanical constitutive equations to a numerical model has also been carried out to analyze deformation behaviors at high temperature. Gomes et al. [39] developed a numerical method to assess the safety design of a steel–timber connection subjected to both mechanical and thermal loadings. Bardel et al. [40] developed a numerical model with consideration of the phase transformation of aluminum alloy 6061 for a welding simulation. In their simulations, a metallurgical model was based on precipitate dissolution and coupled with thermomechanical constitutive equations. Rong et al. [1] developed a numerical model integrated with thermodynamic-based phase transformation to investigate residual stress generation in welding. Deng and Murakawa [41] developed a numerical model for welding simulation considering volume change due to martensite transformation, along with the thermomechanical behaviors of 9Cr-1Mo-alloy steel.

DH36 steel is a material widely used in the shipbuilding industry. It has high strength and good resistance against corrosion and fatigue at low temperatures. In welding, thermal profiles play a crucial role in the evolution of microstructure, and thus residual stress generation. In order to improve the reliability of welded components, it is necessary to understand the underlying mechanism of residual stress generation in welding. Among several welding process parameters, welding speed has a great influence on thermal profile and residual stress. Though several works have reported the effects of welding speed for different materials, the influence of welding speed on residual stress in DH36 steel have not been explicitly studied.

In the present study, a thermo-mechanical model incorporated with the constitutive equations of transformation plasticity was developed for welding simulations. In order to reflect the phase transformation during welding, the equation of phase evolution proposed by Leblond and Devaux [42] was implemented into the present numerical model, which was calibrated with a CCT diagram for the various cooling rates. The welding speed, arc voltage, and welding current were set to be 25–60 cm/min, 30.9 V, and 1170 A, respectively. The present numerical model allows the multiphysics simulation which is coupled with thermal, mechanical, and metallurgical behaviors of a material during welding. The numerical implementation was conducted using the ABAQUS user subroutines, UMAT, UMATHT, DFLUX, and UEXPAN. Some of the numerical studies take into account martensite transformation during welding (for example, refs. [2,43,44]). However, to author's best knowledge, there are very few numerical studies in the literature on DH36 welding considering the phase transformation and transformation plasticity of all possible phases as the present study does. The results show that the lower welding speed leads to the higher peak temperature and slow cooling rate which result in the ferrite-dominant microstructures, while the phase fraction of martensite increases with the welding speed because of the fast cooling rate. It can be concluded that the welding speed has the significant effects on the thermal profiles and phase transformation, which lead to the generation of residual stress.

2. Methodology

2.1. Thermal-Metallurgical Analysis

In the present study, equations of the phase evolution proposed by Leblond and Devaux [42] were used to consider the phase transformation and transformation plasticity which is expressed as follows:

$$\dot{p}_i = -\sum_{j=1, j\neq i}^{4} A_{ij}(T, \dot{T}), \quad i = 1, 2, \ldots, 4 \tag{1}$$

where

$$\sum_{i=1}^{4} p_i = 1 \tag{2}$$

In Equation (1), p is the phase volume proportion. The subscript i = 1, 2, 3, and 4 represents ferrite, bainite, martensite, and austenite, respectively. $A_{ij}(T, \dot{T})$, refers to the rate of the phase transformed from j to i. Note that $A_{ij}(T, \dot{T})$ is a function of temperature and temperature change rate. It is explained with more details in Section 2.2.

For martensitic transformation, the Koistinen–Marburger model [45] is used as Equation (3):

$$\begin{gathered} p_3(T) = 1 - \exp\left\{a\left(T_{3,S} - T\right)\right\}\left(T \leq T_{3,S}\right) \\ \text{with } a = -\frac{\ln\left[p_1\left(T_{3,S}\right) + p_2\left(T_{3,S}\right)\right]}{T_{3,S} - T_{3,F}} \end{gathered} \tag{3}$$

Here, $T_{3,S}$ and $T_{3,F}$ are the start and finish temperatures for the martensite transformation. The energy balance equation and thermal boundary conditions are coupled together as in Equations (4)–(6). The phase proportions are obtained by solving Equation (1). Such results are plugged into Equation (4) for thermal analysis.

$$\sum_i p_i(\rho c)_i \frac{dT}{dt} + \sum_i \dot{p}_i \rho_i H_i = \nabla \cdot \left(\sum_i p_i \lambda_i \nabla T\right) \tag{4}$$

$$(-)\lambda \frac{\partial T}{\partial n} = q \text{ on } S_q \tag{5}$$

$$(-)\lambda \frac{\partial T}{\partial n} = \chi_1(T - T_0) + \chi_2(T^4 - T_0^4) \text{ on } S_\theta \tag{6}$$

Here, ρ_i is the density; c_i is specific heat capacity; H_i is enthalpy; and λ_i is thermal conductivity. S_q represents the surface where the heat flux is applied; n is the normal vector to the surface; q is the heat flux from the heat source; S_θ represents the surface where the convection and radiation heat transfers occur. T_0 is the ambient temperature; and χ_1 and χ_2 are the convective coefficient and radiative coefficient, respectively.

In this work, the double ellipsoidal heat source model [46] is used for the welding simulations. In the model, the volumetric heat is applied in the form of Gaussian distribution as in Equations (7) and (8).

$$Q_f = \frac{6\sqrt{3}f_f Q}{a_f bc\pi\sqrt{\pi}} \exp\left(-\frac{3x^2}{a_f^2} - \frac{3y^2}{b^2} - \frac{3z^2}{c^2}\right) \tag{7}$$

$$Q_r = \frac{6\sqrt{3}f_r Q}{a_r bc\pi\sqrt{\pi}} \exp\left(-\frac{3x^2}{a_r^2} - \frac{3y^2}{b^2} - \frac{3z^2}{c^2}\right) \tag{8}$$

where a_f and a_r indicate the front and rear lengths of the heat source, respectively; and b and c indicate the width and depth of the heat source, respectively. f_f and f_r are parameters that assign the fraction of the heat applied in the front and rear parts, respectively. Q_f and Q_r are the heat energy of the front and rear parts in the double ellipsoidal heat source model, respectively. Q is the welding heat power. The values for the heat source parameters were chosen from the reference [47] and listed in Table 1.

Table 1. Goldak's heat source parameters (taken from [47]).

a_f (mm)	a_r (mm)	b (mm)	c (mm)	f_f	f_r
20	80	20	22	0.4	1.6

2.2. Calibration with CCT Diagram

In the present study, the calibration with the CCT diagram [48] of DH36 steel was conducted for the precise calculation of the phase volume proportion. The term $A_{ij}(T, \dot{T})$ can be re-expressed as follows:

$$A_{ij} = \begin{cases} k_{ij}(T)p_i - l_{ij}(T)p_j \\ \text{if } k_{ij}(T)p_i - l_{ij}(T)p_j > 0 (\text{from i to j transf.}) \\ -k_{ji}(T)p_j + l_{ji}(T)p_i \\ \text{if } k_{ji}(T)p_j - l_{ji}(T)p_i > 0 (\text{from j to i transf.}) \\ 0 \text{ if } k_{ij}(T)p_i - l_{ij}(T)p_j \le 0 \\ \text{and } k_{ji}(T)p_j - l_{ji}(T)p_i \le 0 \\ (\text{no transformation between phases i and j}) \end{cases} \tag{9}$$

where k_{ij} and l_{ij} are the parameters to be determined from the calibration with the CCT diagram. Following the suggestion in the reference of Leblond and Devaux [42], they are defined as follows:

$$k_{ij} = \frac{P_{eq}(T)}{\tau}, l_{ij} = \frac{1 - P_{eq}(T)}{\tau} \tag{10}$$

where $P_{eq}(T)$ is the equilibrium phase volume proportion after an infinitely long time, which has a value between 0 and 1. τ is the characteristic time necessary for an equilibrium state. $P_{eq}(T)$ is determined by Equation (11) as follows:

$$P_{eq}(T) = \frac{T_s - T}{T_s - T_f} \tag{11}$$

Note that the derivative of the phase can be defined as Equation (12)

$$\frac{dP}{dt} = \frac{P_{eq} - P}{\tau} \tag{12}$$

From Taylor's expansion to Equation (12), we have the phases (P_{n+1}) at the time t_{n+1} as in Equation (13):

$$P_{n+1} = P_n + \left.\frac{dP}{dt}\right|_n \left(\frac{dT}{dt}\right)^{-1} \Delta T = P_n + \frac{P_{eq} - P}{\tau} \left(\frac{dT}{dt}\right)^{-1} \Delta T \tag{13}$$

Calibration of the CCT diagram of DH36 steel was conducted for metallurgical analysis. In the present study, the CCT diagrams in the reference [48] were used for the calibration. Iterative computations of Equation (13) were performed by finding τ such that the temperature T_f was in good agreement with the one in the CCT diagram. Table 2 shows the metallurgical parameters, τ determined from the calibration for ferritic and bainitic transformations. Tables 3–5 show the start and finish temperatures for ferritic and bainitic transformations, which were obtained from the present study and the CCT diagram. The results show that the metallurgical analysis in the present study is in good agreement with the one in CCT diagram. The maximum differences in temperature from the two results are overall less than 5 °C.

Table 2. The parameter (τ) for metallurgical analysis in the present numerical model.

Cooling Rate [°C/s]	Peak Temperature [°C]	τ [s] (Ferritic Transformation)	τ [s] (Bainitic Transformation)
1	875	0.92	0.0002
1	1000	0.94	0.0004
1	1350	0.001	0.005
5	875	0.02	0.012
5	1150	0.003	0.004
5	1350	0.96	0.0231
10	875	1.05	0.0154
10	1350	1.26	0.024
25	875	1.45	0.0369
25	1350	1.654	0.005
100	875	1.032	0.005
100	1000	1.727	0.027
200	875	1.234	0.0031
200	1000	1.256	0.0284

Table 3. Comparison of the start temperature calculations for ferritic transformation from the present simulations with the ones in the CCT diagram as a function of cooling rate and peak temperature.

Cooling Rate [°C/s]	Peak Temperature [°C]	CCT [°C]	Present Result [°C]
1	875	787	783
1	1000	788	785
1	1150	752	754
1	1350	768	771
5	875	755	752
5	1000	750	750
5	1150	700	703
5	1350	689	692
10	875	762	759
10	1000	744	747
10	1150	694	692
10	1350	660	665

Table 3. *Cont.*

Cooling Rate [°C/s]	Peak Temperature [°C]	CCT [°C]	Present Result [°C]
25	875	788	782
25	1000	717	715
25	1150	682	680
25	1350	627	620
100	875	678	673
100	1000	704	708
200	875	682	685
200	1000	689	692

Table 4. Comparison of the finish temperature calculations for ferritic transformation from the present simulations with the ones in the CCT diagram as a function of cooling rate and peak temperature.

Cooling Rate [°C/s]	Peak Temperature [°C]	CCT [°C]	Present Result [°C]
1	875	674	669
1	1000	672	674
1	1150	596	592
1	1350	572	578
5	875	657	655
5	1000	658	654
5	1150	545	550
5	1350	564	559
10	875	643	642
10	1000	649	643
10	1150	546	542
10	1350	521	520
25	875	607	609
25	1000	585	592

Table 5. Comparison of the start temperature calculations for bainitic transformation from the present simulations with the ones in the CCT diagram as a function of cooling rate and peak temperature.

Cooling Rate [°C/s]	Peak Temperature [°C]	CCT [°C]	Present Result [°C]
100	875	674	669
100	1000	672	674
100	1150	596	592
200	1350	572	578
200	875	657	655
200	1000	658	654

2.3. Mechanical Analysis

A thermo-mechanical constitutive model incorporated with the transformation plasticity proposed by Leblond et al. [11,12] were implemented into the present numerical model. The multiple phases in steel were categorized into two phases: the hard phase (α-phase: ferrite, bainite, and martensite) and the weak phase (γ-phase: the austenite). As discussed above, Fe atoms in steel are arranged in either BCC or FCC structure depending on temperature. Ferritic steel has a BCC structure at room temperature, while austenite has a FCC structure above the eutectoid temperature. The terms of "hard" and "weak" are originated from the level of yield stress. The hard phase has the higher yield stress than the weak phase.

The total strain ε^{tot} is composed of the elastic strain ε^{el}, the thermal strain ε^{th}, the strain due to transformation plasticity ε^{tp}, and the strain due to conventional plasticity ε^{cp}, as shown in Equation (14).

$$\varepsilon^{tot} = \varepsilon^{el} + \varepsilon^{th} + \varepsilon^{tp} + \varepsilon^{cp} \quad (14)$$

The yield stress of the α-phase (σ_2^y) is calculated as an average of yield stresses of each phase as shown in Equation (15):

$$\sigma_2^y(T) = \sum_i p_i \sigma_i^y \tag{15}$$

where p_i denotes the volume proportion of each phase, while σ_i^y is the yield stress of each phase i. The mixture rule is applied to determine the yield stress of multiple-phases and is given by:

$$\sigma^y\left(\varepsilon_1^{eff}, \varepsilon_2^{eff}, T\right) = [1 - f(z)]\sigma_1^y\left(T, \varepsilon_1^{eff}\right) + f(z)\sigma_2^y\left(T, \varepsilon_2^{eff}\right) \tag{16}$$

Here, σ_1^y and σ_2^y, is the yield stress of the weak phase and the hard phase, respectively. z is a summation of the phase volume fraction of the hard phases. $f(z)$ is the modification factor [12] for the nonlinear mixture rule listed in Table 6.

Table 6. Values of functions f(z) and g(z) (taken from [12]).

z	f(z)	g(z)
0	0	0
0.125	0.044	2.53
0.25	0.125	4
0.5	0.391	2.76
0.75	0.668	1.33
1.0	1	1

Following the Leblond's characterization of the flow rule [11,12], the plastic deformation in the present numerical model is categorized into the two types: the transformation plasticity and the conventional macroscopic plasticity.

First, the deformation is assumed to be a purely elastic and the trial stress is calculated as follows:

$$\sigma_{n+1}^{trial} = \sigma_n + C^{el} : \Delta\hat{\varepsilon}^{RN} + \Delta C : \varepsilon_n^{el} \tag{17}$$

where σ_{n+1}^{trial} is the trial stress at the current time step; C^{el} is the elastic tangent moduli; $\Delta\hat{\varepsilon}^{RN}$ is the rotation-neutralized strain increment; and ε_n^{el} is the elastic strain at the previous time step. Once the trial stress is updated, it is compared with the yield stress to determine whether a material undergoes the elastic or plastic deformation. If the deformation turns out to be elastic, the final stress is updated with the trial stress and the simulation moves to the next time step.

The plastic deformation is categorized into two types as described above: the transformation plasticity ($\overline{\sigma} \leq \sigma^y$, in cooling) and the conventional macroscopic plasticity. For the stress-update of the plastic deformation, the stress relaxation is conducted using the radial-return scheme in a different way, depending on the type of plasticity (Equations (18)–(22) for the transformation plasticity; Equations (23)–(26) for the conventional macroscopic plasticity). The overall schematic of the present numerical model is demonstrated in Figure 1.

Case 1: Transformation plasticity ($\overline{\sigma} \leq \sigma^y$)

$$\dot{\varepsilon}^p = \frac{3}{2}\frac{\dot{\overline{\varepsilon}}^p}{\overline{\sigma}}\mathbf{s} = \sqrt{\frac{3}{2}}\mathbf{n} \text{ with } \mathbf{n} = \sqrt{\frac{3}{2}}\frac{\mathbf{s}}{\overline{\sigma}} \tag{18}$$

$$\dot{\overline{\varepsilon}}^p = \sqrt{\frac{3}{2}}\frac{1-z}{\sigma_1^y\left(\overline{\varepsilon}_1^{eff}\right)}\dot{\varepsilon}_1^{eff} \parallel \mathbf{s} \parallel \tag{19}$$

$$\dot{\bar{\varepsilon}}_1^{\text{eff}} = \frac{-2\Delta\varepsilon_{1\to 2}^{\text{th}}}{1-z}(\ln z)\dot{z}\hat{h}\left(\| \mathbf{s} \|, \dot{\bar{\varepsilon}}_1^{\text{eff}}, \dot{\bar{\varepsilon}}_2^{\text{eff}}\right) + \sqrt{\frac{3}{2}}\frac{g(z)}{E}\| \mathbf{s} \| + \frac{2(\alpha_1-\alpha_2)z\ln z}{1-z}\dot{T} \quad (20)$$

$$\dot{\bar{\varepsilon}}_2^{\text{eff}} = -\frac{\dot{z}}{z}\bar{\varepsilon}_2^{\text{eff}} + \omega^{\dot{z}}\bar{\varepsilon}_1^{\text{eff}} \quad (21)$$

$$\bar{\sigma} = \sqrt{\frac{3}{2}\mathbf{s}:\mathbf{s}} \quad (22)$$

where $\mathbf{s}$ is the deviatoric stress; $\Delta\varepsilon_{1\to 2}^{\text{th}}$ is the difference in thermal strain between the α- and β-phases; $\hat{h}\left(\| \mathbf{s} \|, \dot{\bar{\varepsilon}}_1^{\text{eff}}, \dot{\bar{\varepsilon}}_2^{\text{eff}}\right)$ is a function that takes into account nonlinearities in the stress; E is the Young's modulus; α_1 and α_2 are the coefficients of thermal expansion of the weak and hard phases, respectively. g(z) (listed in Table 6) is a modification function for small values of z. Leblond and Devaux [12] obtained the values of g(z) from the curve fitting on the stress–strain data. Such values are used in the present study and listed in Table 6.

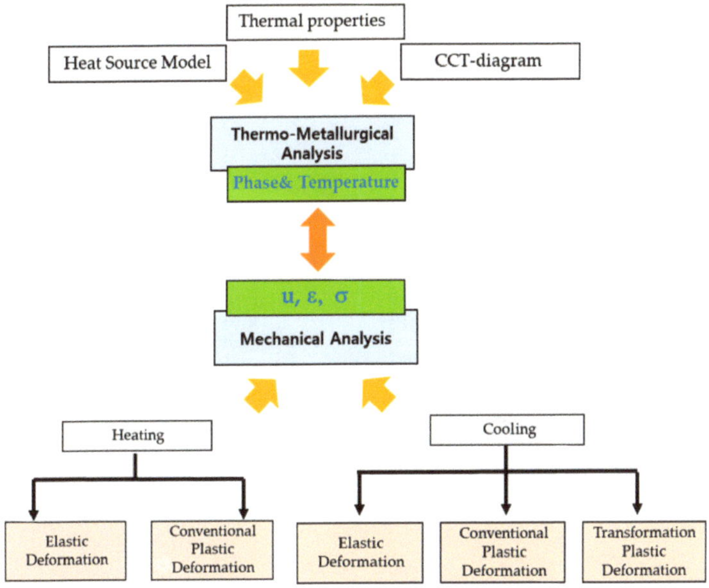

Figure 1. The schematic of the present numerical model.

Case 2: Conventional macroscopic plasticity ($\bar{\sigma} = \sigma^y$)

$$\dot{\varepsilon}^p = \frac{3}{2}\frac{\dot{\bar{\varepsilon}}^p}{\bar{\sigma}}\mathrm{s} \quad (23)$$

$$\bar{\sigma} = \sigma^y \quad (24)$$

$$\dot{\bar{\varepsilon}}_1^{\text{eff}} = \dot{\bar{\varepsilon}}^p \quad (25)$$

$$\dot{\bar{\varepsilon}}_2^{\text{eff}} = \dot{\bar{\varepsilon}}_2^p - \frac{\dot{z}}{z}\bar{\varepsilon}_2^{\text{eff}} + \omega\frac{\dot{z}}{z}\bar{\varepsilon}_1^{\text{eff}} \quad (26)$$

2.4. Finite Element Model

Figure 2 shows the material properties used in the present study. The values of the material properties were taken from [47] and plotted in Figure 2. The yield stresses of each phase are listed in Table 7. The values of thermal expansion coefficients of the weak and the hard phase ($\varepsilon_1 = 2.260 \times 10^{-5}\ ^\circ C^{-1}$; $\varepsilon_2 = 1.615 \times 10^{-5}\ ^\circ C^{-1}$) and the Poisson's ratio (ν = 0.3) were employed from [48]. In the present work, four different welding speeds (v = 25, 35, 40, 60 cm/min) are considered to investigate the influence of welding speed on the residual stress generation. The simulations were performed with and without consideration of the phase transformation for each welding speed. Therefore, total eight simulation cases were conducted for analysis of thermal profiles and residual stress. The simulation cases considered in the present study are listed in Table 8.

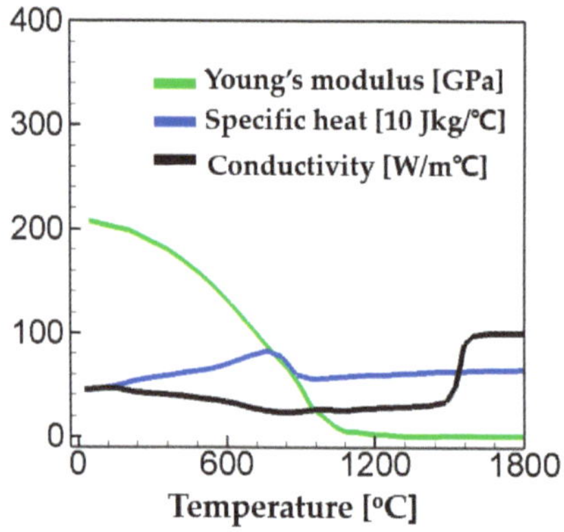

Figure 2. Material properties used in the present study (adapted from [47]).

Table 7. Yield stresses of each phase (taken from [48]).

Temperature [°C]	Ferrite [MPa]	Bainite [MPa]	Martensite [MPa]	Austenite [MPa]
20	425	405	750	220
700	110	100	210	98
1300	12.5	12.5	12.5	12.5

Table 8. Simulation cases considered in the present study.

Case No.	Welding Speed [cm/min]	Consideration of Phase Transformation
Case 1-SP	25	No
Case 1-MP	25	Yes
Case 2-SP	35	No
Case 2-MP	35	Yes
Case 3-SP	40	No
Case 3-MP	40	Yes
Case 4-SP	60	No
Case 4-MP	60	Yes

The numerical model discussed in Sections 2.1–2.3 were implemented into the ABAQUS user-subroutines UMAT, UMATHT, DFLUX, and UEXPAN. In the UEXPAN, the thermal strain of the multiple-phase steel is calculated. In DFLUX, the heat source is defined including the size, position, velocity, and the level of heat energy. The thermal behaviors of a material with an account of the phase transformation are defined in the UMATHT. The calculation of the phase volume fraction and the stress-update are conducted in the UMAT subroutine.

In this work, the dimensions of the FE model are 1100 mm × 600 mm × 12.5 mm. A half symmetry condition was imposed to reduce the computation time, and some of nodes were constrained to prevent the rigid body motion as shown in Figure 3b. The local mesh refinement was applied to the fusion zone (FZ) and heat-affected zone (HAZ) with their minimum size of about 1 mm to optimize the accuracy and efficiency of the FE simulation. Such meshing was determined based on the preliminary simulations which were performed to identify the region where mesh refinement is required. The model was discretized with 58,264 C3D8T elements (8-node trilinear element for thermo-mechanically fully coupled analysis). There could be other options for the choice of element, for example, (1) sequentially coupled elements, or (2) 2D elements. However, C3D8T element was chosen in the present study because it gives the better results with reasonable computation time than other options. The welding conditions in this work is 30.9 V, 1170A for the arc voltage and welding current, respectively. The total computation time was set to 4000 s for the model to reach the room temperature after welding. The time step was chosen as the automatic time increment so that the size of the time step can be adjustable based on the solution convergence rates.

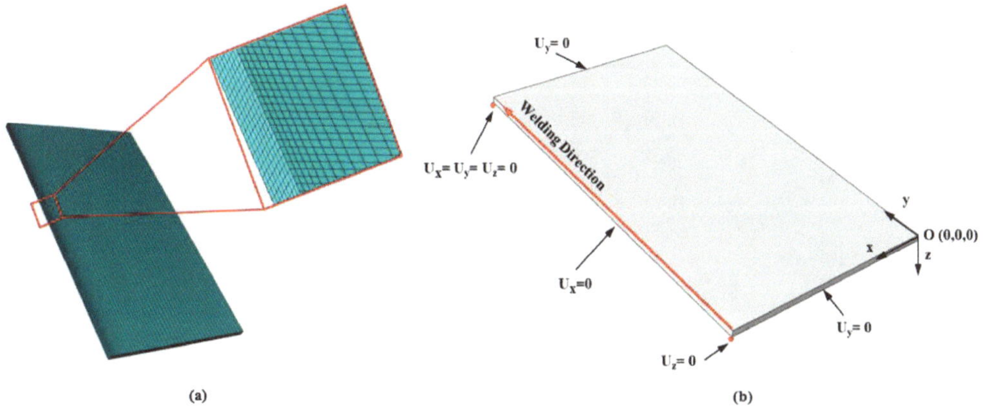

Figure 3. (**a**) Finite element model; (**b**) boundary conditions for the present numerical study.

3. Results and Discussion

First, for verification of thermal analysis in the present numerical model, the values of temperature calculated from the simulations are compared with thermocouple measurements from Fu et al. [47]. Following the experiment set-up, welding speed, welding current, and voltage are set to be 560 A, 30 V, and 27 cm/min, and three points of investigation are selected. The comparison results are plotted in Figure 4 and listed in Table 9. As shown, the temperature history curves from the measurements and the simulations are good agreement with each other with the maximum error of 21.6%.

Next, the effects of welding speed on thermal profiles are investigated for the cases listed in Table 8. For such an investigation, the point A was chosen, which was located 8 mm away from the center of the welding line, as shown in Figure 5. Hereafter, this point is named as point A. Figure 5 shows the temperature evolution for the different welding speed and the location of the point A. As shown in Figure 5, the peak temperature decreases as welding speed increases. While the peak temperature for case 1 is calculated as T = 1391 °C, the values for cases 2, 3, 4 are T = 1044 °C, T = 764 °C, T = 596 °C, respectively. In case of the cooling rate from the peak temperature to T = 300 °C, it is found that case 4 (30.3 °C/s) has the highest cooling rate, while case 1 has the lowest one (12.1 °C/s) among the cases considered. The results suggest that the slow welding speed allows more heat accumulation and leads to a lower cooling rate.

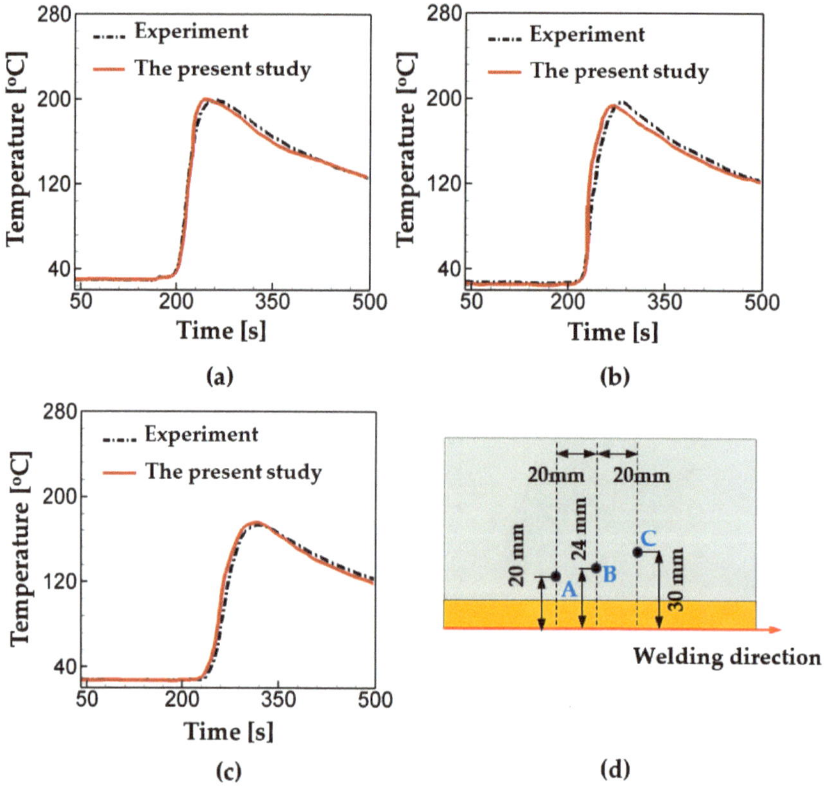

Figure 4. Comparison of the temperature calculations from the present model with the thermocouple measurements at the different locations: (**a**) Point A, (**b**) Point B, (**c**) Point C, and (**d**) the locations of thermocouple arrangement.

Table 9. Comparison of the temperature calculations from the present model with the thermocouple measurements at the different locations: The points of investigation are illustrated in Figure 4d.

Time [s]	Point A			Point B			Point B		
	Num [°C]	Exp [°C]	Error [%]	Num [°C]	Exp [°C]	Error [%]	Num [°C]	Exp [°C]	Error [%]
200	38.2	38.5	0.8	25.6	26.9	5.1	27.6	27.4	1.0
220	126.1	133.3	5.7	30.0	30.2	0.8	27.8	28.1	1.0
240	199.4	191.2	4.1	145.8	114.3	21.6	40.3	33.7	16.4
260	197.6	199.7	1.1	191.7	179.2	6.5	96.5	78.4	18.7
280	191.7	194.5	1.5	188.5	197.1	4.6	151.1	138.4	8.4
300	183.1	186.2	1.7	181.3	189.0	4.3	173.6	168.5	2.9
320	170.4	177.0	3.9	172.0	180.5	4.9	175.2	173.4	1.0
340	161.8	165.5	2.3	165.8	172.3	3.9	167.9	169.1	0.7
360	155.9	162.1	4.0	156.3	161.2	3.1	159.8	160.8	0.7
380	149.1	153.0	2.6	146.7	152.8	4.2	150.5	154.1	2.4
400	145.6	148.3	1.9	144.5	148.9	3.1	142.7	147.2	3.1
420	142.2	141.9	0.2	137.7	142.6	3.6	136.4	140.0	2.6
440	139.2	139.8	0.4	131.1	134.6	2.7	132.6	135.2	2.1
460	134.7	133.0	1.2	127.8	130.5	2.1	125.3	130.6	4.2
480	130.5	129.9	0.5	124.7	125.8	0.9	119.0	127.4	7.1
500	125.3	125.4	0.1	122.0	123.1	0.9	118.2	123.2	3.6

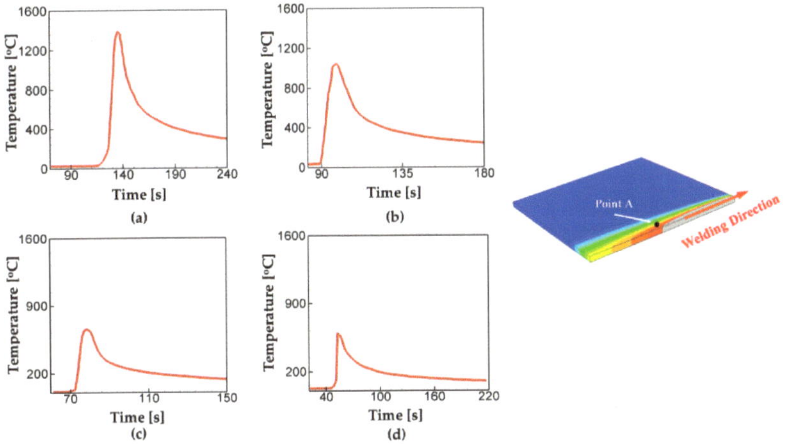

Figure 5. Temperature evolution at the point A for the different welding speed: (**a**) case 1, (**b**) case 2, (**c**) case 3, and (**d**) case 4. The point A locates 8 mm away from the welding line. Details on the welding conditions are listed in Table 8.

Figure 6 shows the phase fraction at the Point A for different welding speed. The evolution of the microstructures in welding is significantly affected by thermal profiles. As described above, iron atoms in any steel are in a BCC structure at room temperature. When temperature exceeds the eutectoid temperature in heating process, the ferritic steel starts to transform into the austenitic steel, having a FCC atomistic structure. In cooling, austenite transforms into the ferrite, bainite, and martensite, depending on the thermal profiles. Analysis of Figures 5 and 6 shows that the time for temperature to reach the eutectoid temperature (T = 741 °C [32]) coincides with the ones when austenite starts to be formed in heating process for cases 1~3. As shown in Figure 5, temperature history reaches the eutectoid temperature at t = 130 s, t = 96 s, and t = 75 s for cases 1~3. Such values are in overall good agreement with the times when phase transformation starts for each case as shown in Figure 6. Note that in Figure 5d, the peak temperature for case 4 is calculated as T = 596 °C, which means it does not pass the critical eutectoid temperature. This explains why phase transformation does not occur at the Point A for case 4 as shown in Figure 6d. In cooling, austenite starts to transform back into ferritic steel depending on thermal profiles such as the peak temperature, cooling rate, and temperature range. For case 1, the final phase volume fractions of ferrite and bainite are calculated as $p_1 = 0.51$ and $p_2 = 0.38$, respectively. For case 2, the values of ferrite, bainite, martensite are calculated as $p_1 = 0.26$, $p_2 = 0.44$ and $p_3 = 0.30$, respectively. During the cooling process of welding, carbon atoms hardly diffuse out of the ferrous cell and trapped in the cell. Therefore, they are distorted and form a martensite phase. Differently from case 1, the cooling rates of case 2 (i.e., 24.2 °C/s) and case 3 (i.e., 28.3 °C/s) are high enough to nucleate and grow martensite phase, leading to the final phase fraction of $p_3 = 0.30$ for case 2 and $p_3 = 0.48$ for case 3. The results show that the phase fraction of martensite increases with an increase in cooling rate. Such a finding can be also found in other research (for example, Ravikumar et al. [49]). Figure 6e shows the phase fraction contours of ferrite and bainite at the end of the simulation for case 1, showing the heat-affected zone (HAZ) near the welding line. The phase fractions in Figure 6 show the consistency with other microstructure analyses of DH36 welding (for example, refs. [50,51]). Toumpis et al. [50] reported that the slow speed welding (10 cm/min) resulted in ferrite-dominant and homogeneous microstructures with refined grains; in the case of the intermediate welding speed (25 cm/mm), it was found that the microstructure consisted of acicular shaped bainite regions and ferrite-dominant regions; For the fast welding (50 cm/min), heterogeneous microstructures were exhibited with predominant bainite. It was reported that the phase fractions of bainite increases considerably with the cooling rate. Reynolds et al. [51] reported that bainitic and martensitic microstructures were formed at the welding speed of 45 cm/min.

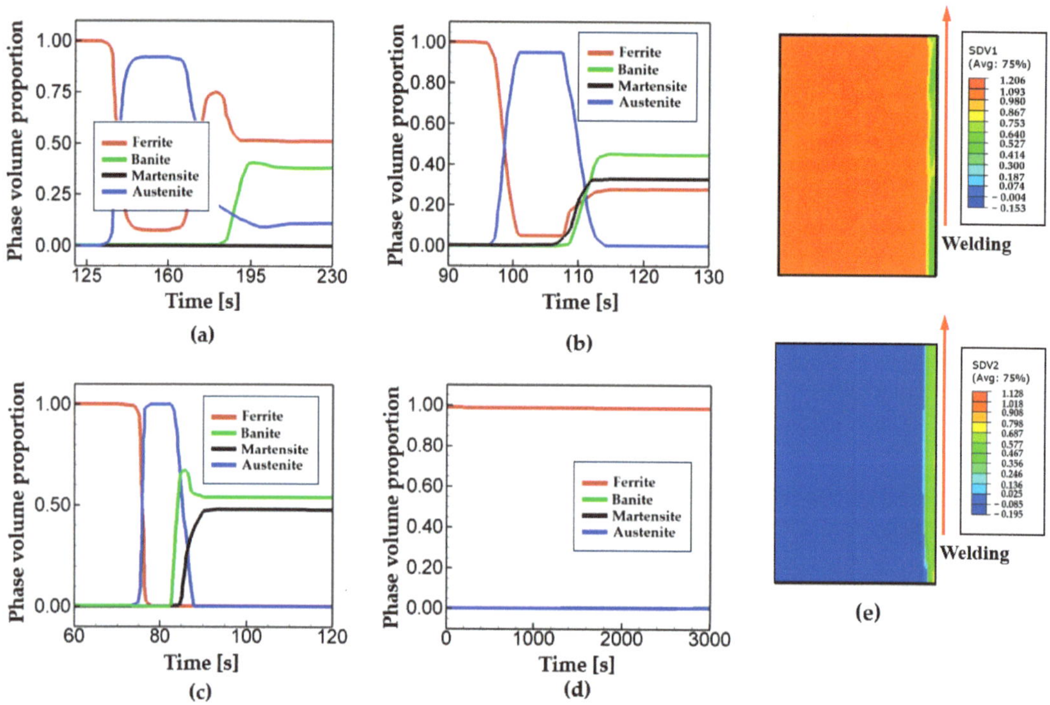

Figure 6. Phase evolution at the point A for the different welding speed: (**a**) case 1, (**b**) case 2, (**c**) case 3, (**d**) case 4, and (**e**) the phase fraction contours of ferrite and bainite at the end of the simulation for case 1.

In this work, the effects of welding speed on thermal profiles, and residual stress are investigated. For such investigation, simulations were performed with and without considering the phase transformation for the different welding speed. All cases considered in the present study are listed in Table 8. In Table 8, 'sp' (single phase) cases indicate the simulations neglecting phase transformation, and thus, the elastic and conventional plasticity are only considered. On the contrary, the 'mp' (multiple phases) cases are the simulations considering the phase transformation and transformation plasticity. In these cases, the elastic, conventional plastic, and transformation plastic deformation are considered as described in Section 2.3.

Figure 7 shows the Mises stress history at the Point A for the different welding speed. First, analyses show that the residual stress increase with a decrease in the welding speed. This is mainly because the welding with lower speed allows more time for heat accumulation to the weldment. This eventually leads to the higher peak temperature and slow cooling rate which result in the ferrite-dominant microstructures. It is worth noting in Figures 5 and 6 that the phase portion of martensite increase with the welding speed due to its high cooling rate. These findings support the trend of Mises stress curves in Figure 7.

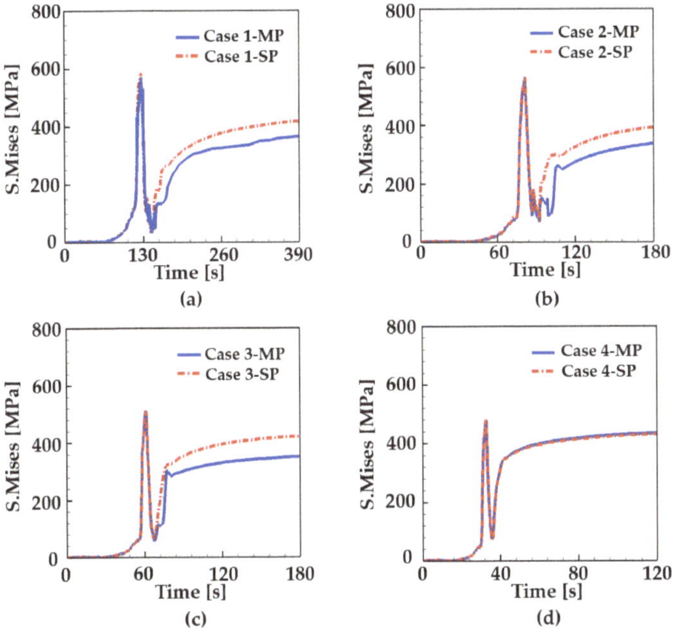

Figure 7. Mises stress as a function of time for different welding speed at the point A. (**a**) case 1, (**b**) case 2, (**c**) case 3, and (**d**) case 4. Refer to Table 8 for details on the simulation conditions.

Another interesting result of Figure 7 is that there exists a clear difference in Mises stress between the sp and mp cases. In other words, the values of Mises stress for the mp cases are lower than the ones for the sp cases for each welding speed. Such a difference results from the volume dilation between coexisting phases of steel. In the heating process, a compressive stress is generated in the FZ and HAZ. As heat continues to be applied to the weldment, the compressive stress gets smaller because the yield stress decreases at high temperatures. When heating is completed and cooling starts, the tensile stress is generated because of the shrinkage of the weldment [44]. However, when temperature reaches in the range of phase transformation (787–545 °C), the compressive stress is generated again due to the volume difference between the coexisting phases (i.e., transformation plasticity). Once the phase transformation is completed and the phase volume fraction is finalized, the amount of compressive stress produced due to multiple phases also remains constant. In Figure 7, the difference in Mises stress between the sp and mp cases for each welding speed becomes constant at t = 210 s, t = 115 s, and t = 90 s for case 1, 2, and 3, respectively. Such values of time are good agreement with the ones when the phase transformation is completed and the phase volume fraction is finalized in Figure 6. The same trends were also observed in other studies in the literatures. For example, Cho and Kim [52] performed the simulations and experiments of welding to investigate the effects of martensite transformation on the residual stress generation in AISI 1045 and AISI 1020 steels. In their experiments, the compressive stress was observed in the weld zone, which was produced due to volume change from phase transformation. Tajat et al. [53] measured the welding-induced residual stress using the neutron diffraction technique. The results showed that the longitudinal stress in the FZ and HAZ is lower than the one in base material adjacent to the HAZ. However, it is worth to noting that there exists a variation in thermal expansion at the interface between the HAZ and base metal due to their different microstructural features. Such variation results in the tensile stress near the interface as reported in previous studies (e.g., refs. [54,55]). This is detrimental to the fatigue life of the components. Note that in the case of a welding speed of v = 60 cm/min (i.e., case 4), no phase transformation occurs because its peak temperature does not reach the eutectoid temperature, as shown in Figure 7d. This explains why the Mises stress curves for case 4-mp and case 4-sp coincide each other.

In order to analyze the influence of welding speed on the residual stress generation, the values of Mises stress for the sp and mp cases are compared and plotted as a function of the distance from welding line as shown in Figure 8 along with the yield stress of a welded steel at the point A. Note that the yield stress is various depending on the cooling rate. The point A was chosen for the comparison. In Figure 8, the zone where a clear difference between the sp and mp cases is observed is the HAZ. Figure 8 shows that the width of HAZ decreases with an increase in welding speed, and thus, the effect of phase transformation on the residual stress generation in a weldment is also reduced. For case 1 (welding speed = 25 cm/min), the width of HAZ is calculated as 66 mm, while it decreases to 18 mm in case of the welding speed = 40 cm/min (i.e., case 3). If the welding speed is greater than 60 cm/min, the width of HAZ becomes very narrow and there is little effect of phase transformation on residual stress, as shown in Figure 8d.

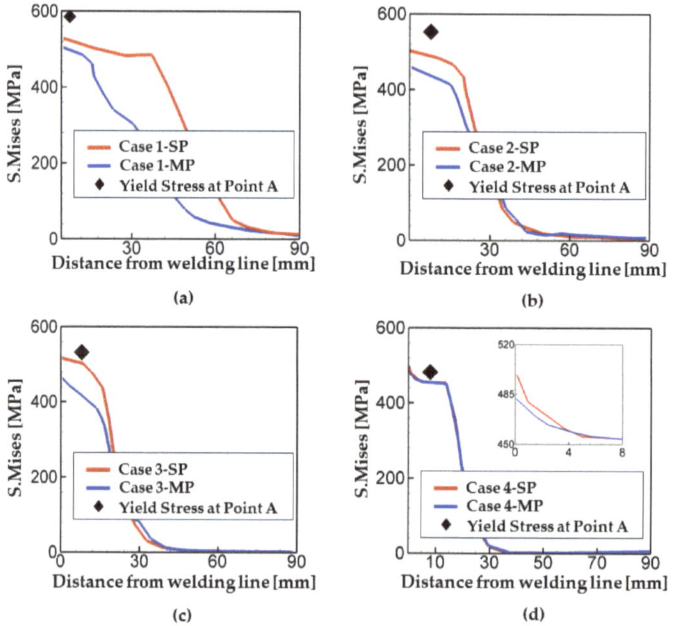

Figure 8. Mises stress as a function of distance from the welding line for different welding speed. (**a**) case 1, (**b**) case 2, (**c**) case 3, and (**d**) case 4. Refer to Figure 5 for the location of point A.

4. Conclusions

In this work, a thermo-mechanical constitutive model considering the phase transformation and transformation plasticity was implemented into the numerical model by way of ABAQUS user-subroutines. In order to consider the phase evolution in welding, the metallurgical parameters for the Leblond's phase equation were obtained from the calibration with a CCT diagram of DH36 steel. In addition, the effects of welding speed on the thermal profiles and residual stress generation were investigated. Major findings are the following:

1. The results show that residual stress increases with a decrease in welding speed. Lower speed of welding allows more time for heat accumulation to a weldment. This leads to the higher peak temperature and slow cooling rate which result in the ferrite-dominant microstructures.

2. The width of HAZ decreases as the welding speed increases. It is shown from the numerical results with and without consideration of phase transformation that the phase volume fraction is significantly affected by the welding speed. Such phase transformation leads to the generation of a compressive stress in the FZ and HAZ.
3. Simulation results show that the phase fraction of martensite increases with a welding speed because of its high cooling rate. The volume difference between the coexisting phases produces a compressive stress in cooling and its magnitude increases with an increase in martensite phase fraction.
4. Since the thermal profiles such as cooling rate and peak temperature are significantly affected by the welding speed, phase transformation should be considered for accurate analysis of residual stress in a weldment.

Residual stress generation is one of the major concerns in welding. Such a challenge mainly results from the rapid heating and cooling during welding. The present numerical model allows the multiphysics simulation which is coupled with thermal, mechanical, and metallurgical behaviors of a material. The present work can be extended to the welding of different materials such as aluminum and titanium alloys. An investigation on the welding process parameters such as interpass idle time and welding trajectory for multipass welding is underway to establish the optimal welding conditions.

Funding: This research received no external funding.

Institutional Review Board Statement: Not applicable.

Informed Consent Statement: Not applicable.

Data Availability Statement: The data that support the findings of this study are available from the corresponding author upon request.

Conflicts of Interest: The author declares no conflicts of interest.

References

1. Rong, Y.; Lei, T.; Xu, J.; Huang, Y.; Wang, C. Residual stress modelling in laser welding marine steel EH36 considering a thermodynamics-based solid phase transformation. *Int. J. Mech. Sci.* **2018**, *146–147*, 180–190. [CrossRef]
2. Lee, C.-H.; Chang, K.-H. Prediction of residual stresses in high strength carbon steel pipe weld considering solid-state phase transformation effects. *Comput. Struct.* **2011**, *89*, 256–265. [CrossRef]
3. Ahmed, M.M.Z.; El-Sayed Seleman, M.M.; Elfishawy, E.; Alzahrani, B.; Touileb, K.; Habba, M.I.A. The Effect of Temper Condition and Feeding Speed on the Additive Manufacturing of AA2011 Parts Using Friction Stir Deposition. *Materials* **2021**, *14*, 6396. [CrossRef]
4. Kenel, C.; Grolimund, D.; Li, X.; Panepucci, E.; Samson, V.A.; Sanchez, D.F.; Marone, F.; Leinenbach, C. In situ investigation of phase transformations in Ti-6Al-4V under additive manufacturing conditions combining laser melting and high-speed micro-X-ray diffraction. *Sci. Rep.* **2017**, *7*, 16358. [CrossRef]
5. Mitra, S.; Arora, K.S.; Bhattacharya, B.; Singh, S.B. Effect of Welding Speed on the Prediction Accuracy of Residual Stress in Laser Welded 1.2 mm Thick Dual Phase Steel. *Lasers Manuf. Mater. Process.* **2020**, *7*, 74–87. [CrossRef]
6. Doumenc, G.; Couturier, L.; Courant, B.; Paillard, P.; Benoit, A.; Gautron, E.; Girault, B.; Pirling, T.; Cabeza, S.; Gloaguen, D. Investigation of microstructure, hardness and residual stresses of wire and arc additive manufactured 6061 aluminium alloy. *Materialia* **2022**, *25*, 101520. [CrossRef]
7. Singh, U.P.; Swaminathan, S.; Phanikumar, G. Thermo-mechanical approach to study the residual stress evolution in part-scale component during laser additive manufacturing of alloy 718. *Mater. Des.* **2022**, *222*, 111048. [CrossRef]
8. Sun, Q.; Liu, Y.B.; Sun, Q.J.; Wang, Y.X. The Effect of Multiple Thermal Process on Microstructural Evolution and Mechanical Properties of Additive Manufactured Al/Steel Structure. *Adv. Eng. Mater.* **2022**, *24*. [CrossRef]
9. Xia, J.; Jin, H. Numerical modeling of coupling thermal–metallurgical transformation phenomena of structural steel in the welding process. *Adv. Eng. Softw.* **2018**, *115*, 66–74. [CrossRef]
10. Tsirkas, S.; Papanikos, P.; Kermanidis, T. Numerical simulation of the laser welding process in butt-joint specimens. *J. Mater. Process. Technol.* **2003**, *134*, 59–69. [CrossRef]
11. Leblond, J.B.; Mottet, G.; Devaux, J.C. A theoretical and numerical approach to the plastic behaviour of steels during phase transformations—I. Derivation of general relations. *J. Mech. Phys. Solids* **1986**, *34*, 395–409. [CrossRef]
12. Leblond, J.B.; Mottet, G.; Devaux, J.C. A theoretical and numerical approach to the plastic behaviour of steels during phase transformations—II. Study of classical plasticity for ideal-plastic phases. *J. Mech. Phys. Solids* **1986**, *34*, 411–432. [CrossRef]

13. Rai, R.; Kelly, S.M.; Martukanitz, R.P.; DebRoy, T. A Convective Heat-Transfer Model for Partial and Full Penetration Keyhole Mode Laser Welding of a Structural Steel. *Metall. Mater. Trans. A* **2008**, *39*, 98–112. [CrossRef]
14. Hammad, A.; Churiaque, C.; Sánchez-Amaya, J.M.; Abdel-Nasser, Y. Experimental and numerical investigation of hybrid laser arc welding process and the influence of welding sequence on the manufacture of stiffened flat panels. *J. Manuf. Process.* **2021**, *61*, 527–538. [CrossRef]
15. Huang, H.; Tsutsumi, S.; Wang, J.; Li, L.; Murakawa, H. High performance computation of residual stress and distortion in laser welded 301L stainless sheets. *Finite Elem. Anal. Des.* **2017**, *135*, 1–10. [CrossRef]
16. Diao, Z.; Yang, F.; Zhang, Y.; Chen, L.; Xiong, T.; Yang, W.; Rong, M. Effect of heat treatment on the microstructure and properties of CuCrZr alloy manufactured by wire arc additive manufacturing. *J. Alloys Compd.* **2023**, *967*, 171786. [CrossRef]
17. Gao, L.; Bhattacharyya, J.; Lin, W.; Ren, Z.; Chuang, A.C.; Shevchenko, P.D.; Nikitin, V.; Ma, J.; Agnew, S.R.; Sun, T. Tailoring material microstructure and property in wire-laser directed energy deposition through a wiggle deposition strategy. *Addit. Manuf.* **2023**. [CrossRef]
18. Hong, X.; Huang, B.; Li, T.; Chen, P.; Zheng, J.; Zhu, Y. Effects of Welding Speed and Welding Current on the Residual Stress and Deformation of SAF 2507/316L Dissimilar Plasma Arc Welding. *J. Mater. Eng. Perform.* **2023**. [CrossRef]
19. Kazmi, K.H.; Sharma, S.K.; Das, A.K.; Mandal, A.; Kumar Shukla, A.; Mandal, R. Wire Arc Additive Manufacturing of ER-4043 Aluminum Alloy: Effect of Tool Speed on Microstructure, Mechanical Properties and Parameter Optimization. *J. Mater. Eng. Perform.* **2023**. [CrossRef]
20. Palmeri, D.; Pollara, G.; Licari, R.; Micari, F. Finite Element Method in L-PBF of Ti-6Al-4V: Influence of Laser Power and Scan Speed on Residual Stress and Part Distortion. *Metals* **2023**, *13*, 1907. [CrossRef]
21. Paolo Ferro, F.B.A.A.T. Methodologies and experimental validations of welding process numerical simulation. *Int. J. Comput. Mater. Sci. Surf. Eng.* **2010**, *3*, 114–132.
22. Gannon, L.; Liu, Y.; Pegg, N.; Smith, M. Effect of welding sequence on residual stress and distortion in flat-bar stiffened plates. *Mar. Struct.* **2010**, *23*, 385–404. [CrossRef]
23. Huang, H.; Wang, J.; Li, L.; Ma, N. Prediction of laser welding induced deformation in thin sheets by efficient numerical modeling. *J. Mater. Process. Technol.* **2016**, *227*, 117–128. [CrossRef]
24. Biswas, P.; Kumar, D.A.; Mandal, N.R.; Mahapatra, M.M. A study on the effect of welding sequence in fabrication of large stiffened plate panels. *J. Mar. Sci. Appl.* **2011**, *10*, 429–436. [CrossRef]
25. Gurmesa, F.D.; Lemu, H.G. Literature Review on Thermomechanical Modelling and Analysis of Residual Stress Effects in Wire Arc Additive Manufacturing. *Metals* **2023**, *13*, 526. [CrossRef]
26. Zhou, G.; Liu, B.; Song, W.; Li, H.; Kuang, J.; Qiu, M. Controlling Welding Residual Stress and Distortion of High-Strength Aluminum Alloy Thin Plates by a Trailing Hybrid High-Speed Gas Fluid Field. *Materials* **2022**, *15*, 6451. [CrossRef] [PubMed]
27. Zhou, Q.; Jiang, P.; Shao, X.; Gao, Z.; Cao, L.; Yue, C.; Li, X. Optimization of Process Parameters of Hybrid Laser–Arc Welding onto 316L Using Ensemble of Metamodels. *Metall. Mater. Trans. B* **2016**, *47*, 2182–2196. [CrossRef]
28. Khoshroyan, A.; Darvazi, A.R. Effects of welding parameters and welding sequence on residual stress and distortion in Al6061-T6 aluminum alloy for T-shaped welded joint. *Trans. Nonferrous Met. Soc. China* **2020**, *30*, 76–89. [CrossRef]
29. Taheri-Behrooz, F.; Aliha Mohammad, R.M.; Maroofi, M.; Hadizadeh, V. Residual stresses measurement in the butt joint welded metals using FSW and TIG methods. *Steel Compos. Struct.* **2018**, *28*, 759–766.
30. Asadi, P.; Alimohammadi, S.; Kohantorabi, O.; Soleymani, A.; Fazli, A. Numerical investigation on the effect of welding speed and heat input on the residual stress of multi-pass TIG welded stainless steel pipe. *Proc. Inst. Mech. Eng. Part B J. Eng. Manuf.* **2020**, *235*, 1007–1021. [CrossRef]
31. Semple, J.K.; Bechetti, D.H.; Zhang, W.; Fisher, C.R. Temperature-Dependent Material Property Databases for Marine Steels—Part 2: HSLA-65. *Integr. Mater. Manuf. Innov.* **2022**, *11*, 13–40. [CrossRef]
32. Semple, J.K.; Bechetti, D.H.; Zhang, W.; Fisher, C.R. Temperature-Dependent Material Property Databases for Marine Steels—Part 3: HSLA-80. *Integr. Mater. Manuf. Innov.* **2022**, *11*, 648–674. [CrossRef]
33. Semple, J.K.; Bechetti, D.H.; Zhang, W.; Fisher, C.R. Temperature-Dependent Material Property Databases for Marine Steels—Part 4: HSLA-100. *Integr. Mater. Manuf. Innov.* **2023**, *12*, 130–156. [CrossRef]
34. Kong, F.; Ma, J.; Kovacevic, R. Numerical and experimental study of thermally induced residual stress in the hybrid laser–GMA welding process. *J. Mater. Process. Technol.* **2011**, *211*, 1102–1111. [CrossRef]
35. Benyounis, K.Y.; Olabi, A.G.; Hashmi, M.S.J. Effect of laser welding parameters on the heat input and weld-bead profile. *J. Mater. Process. Technol.* **2005**, *164–165*, 978–985. [CrossRef]
36. Chuaiphan, W.; Srijaroenpramong, L. Effect of welding speed on microstructures, mechanical properties and corrosion behavior of GTA-welded AISI 201 stainless steel sheets. *J. Mater. Process. Technol.* **2014**, *214*, 402–408. [CrossRef]
37. Zhang, M.; Chen, G.; Zhou, Y.; Liao, S. Optimization of deep penetration laser welding of thick stainless steel with a 10kW fiber laser. *Mater. Des.* **2014**, *53*, 568–576. [CrossRef]
38. Lee, S.J.; Nakamura, H.; Kawahito, Y.; Katayama, S. Effect of welding speed on microstructural and mechanical properties of laser lap weld joints in dissimilar Al and Cu sheets. *Sci. Technol. Weld. Join.* **2014**, *19*, 111–118. [CrossRef]
39. Gomes, C.; Fonseca, E.M.M.; Lopes, H.M. Thermomechanical Analysis of Steel-to-Timber Connections under Fire and the Material Density Effect. *Appl. Sci.* **2022**, *12*, 10516. [CrossRef]

40. Bardel, D.; Nelias, D.; Robin, V.; Pirling, T.; Boulnat, X.; Perez, M. Residual stresses induced by electron beam welding in a 6061 aluminium alloy. *J. Mater. Process. Technol.* **2016**, *235*, 1–12. [CrossRef]
41. Deng, D.; Murakawa, H. Prediction of welding residual stress in multi-pass butt-welded modified 9Cr–1Mo steel pipe considering phase transformation effects. *Comput. Mater. Sci.* **2006**, *37*, 209–219. [CrossRef]
42. Leblond, J.B.; Devaux, J. A new kinetic model for anisothermal metallurgical transformations in steels including effect of austenite grain size. *Acta Metall.* **1984**, *32*, 137–146. [CrossRef]
43. Rong, Y.; Mi, G.; Xu, J.; Huang, Y.; Wang, C. Laser penetration welding of ship steel EH36: A new heat source and application to predict residual stress considering martensite phase transformation. *Mar. Struct.* **2018**, *61*, 256–267. [CrossRef]
44. Deng, D. FEM prediction of welding residual stress and distortion in carbon steel considering phase transformation effects. *Mater. Des.* **2009**, *30*, 359–366. [CrossRef]
45. Koistinen, D.P. A general equation prescribing the extent of the austenite-martensite transformation in pure iron-carbon alloys and plain carbon steels. *Acta Metall.* **1959**, *7*, 59–60. [CrossRef]
46. Goldak, J.; Chakravarti, A.; Bibby, M. A new finite element model for welding heat sources. *Metall. Trans. B* **1984**, *15*, 299–305. [CrossRef]
47. Fu, G.; Gu, J.; Lourenco, M.I.; Duan, M.; Estefen, S.F. Parameter determination of double-ellipsoidal heat source model and its application in the multi-pass welding process. *Ships Offshore Struct.* **2015**, *10*, 204–217. [CrossRef]
48. Bechetti, D.H.; Semple, J.K.; Zhang, W.; Fisher, C.R. Temperature-Dependent Material Property Databases for Marine Steels—Part 1: DH36. *Integr. Mater. Manuf. Innov.* **2020**, *9*, 257–286. [CrossRef]
49. Ravikumar, S.V.; Jha, J.M.; Mohapatra, S.S.; Pal, S.K.; Chakraborty, S. Influence of Ultrafast Cooling on Microstructure and Mechanical Properties of Steel. *Steel Res. Int.* **2013**, *84*, 1157–1170. [CrossRef]
50. Toumpis, A.; Galloway, A.; Cater, S.; McPherson, N. Development of a process envelope for friction stir welding of DH36 steel—A step change. *Mater. Des. (1980–2015)* **2014**, *62*, 64–75. [CrossRef]
51. Reynolds, A.P.; Tang, W.; Posada, M.; Deloach, J. Friction stir welding of DH36 steel. *Sci. Technol. Weld. Join.* **2003**, *8*, 455–460. [CrossRef]
52. Cho, S.H.; Kim, J.W. Analysis of residual stress in carbon steel weldment incorporating phase transformations. *Sci. Technol. Weld. Join.* **2002**, *7*, 212–216. [CrossRef]
53. Taljat, B.; Radhakrishnan, B.; Zacharia, T. Numerical analysis of GTA welding process with emphasis on post-solidification phase transformation effects on residual stresses. *Mater. Sci. Eng. A* **1998**, *246*, 45–54.
54. Gurova, T.; da Zona Oeste, C.U.E.; Castello, X.; do Brasil, H.; Estefen, S.F. Analysis of residual stress state in welded steel plates by X-ray diffraction method. *Rev. Adv. Mater. Sci* **2009**, *19*, 172–175.
55. Nasir, N.S.M.; Razab, M.K.A.A.; Mamat, S.; Iqbal, M. Review on welding residual stress. *Stress* **2006**, *2*, 8–10.

materials

Article

Comparison of Microstructure and Mechanical Properties of Ultra-Narrow Gap Metal Active Gas Arc Welded and Narrow Gap Submerged Arc Welded Q235A Low Carbon Steel

Shang Wu [1], Wenkai Xiao [1,*], Lingfei Gong [1] and Fuju Zhang [2]

1 School of Power and Mechanical Engineering, Wuhan University, Wuhan 430072, China; ws13526073335@163.com (S.W.); glf17396155202@163.com (L.G.)
2 Wuhan Narogap Intelligent Equipment Co., Ltd., Wuhan 430072, China; fjzhang@whu.edu.cn
* Correspondence: xiaowenkai@whu.edu.cn; Tel.: +86-139-8616-3866

Abstract: The 18 mm thick Q235A low carbon steel plates were welded via the ultra-narrow gap metal active gas arc welding (ultra-NGMAGW) and narrow gap submerged arc welding (NGSAW), and the microstructure and mechanical properties of the welded joints' area were characterized. The results showed that there is acicular ferrite (AF) in the weld zone of the joint obtained via the ultra-NGMAGW. The AF grains are fine and have a great difference in growth direction, resulting in high local dislocation density. However, there is no AF in the welded joint obtained via the NGSAW. Using numerical simulation analysis of the temperature field distribution and the thermal cycle curve in the welding process of the ultra-NGMAGW, it was found that the mechanism of microstructure evolution is that during the welding process of the ultra-NGMAGW, the heat input is low, the cooling rate is quick, and the residence time in the high temperature region is short. Therefore, pearlite with coarse grains is basically not formed. AF nucleates in different directions with inclusions as the core. The tensile strength of the weld joint obtained via the ultra-NGMAGW is 643 MPa, which corresponds to 139% of that of the base metal, and 132% of that obtained via the NGSAW. The ultra-NGMAGW joints exhibited better tensile strength and higher microhardness than the NGSAW joints, which is mainly due to the existence of AF.

Keywords: ultra-NGMAGW; NGSAW; AF; microstructure evolution

1. Introduction

Low carbon steel refers to steel with a carbon content between 0.05 and 0.25%, which has low strength and hardness, good plasticity, and toughness. Due to its good processability and cost effectiveness, low carbon steel is used in a wide variety of applications such as manufacturing, construction, automotive, and energy industries. In these industries, it is often necessary to weld low carbon steel thick plates. In traditional welding technology, such as submerged arc welding (SAW), it owns a high welding cladding rate and a convenient welding process, but the removal of welding slag is difficult, and for the large welding thermal deformation in SAW, the heat treatment process must be increased to eliminate it after welding. At the same time, the heat input in the welding process is large, which may promote grain coarsening, resulting in the decline of the welded joints. And, Li et al. [1] found that in SAW, the heat-affected zone (HAZ) is usually brittle and weak because the coarse austenite grains and poor microstructure are formed in the welding process.

Narrow gap welding (NGW) is a process to complete multi-layer and multi-pass welds in a much narrower gap than the conventional welding groove width. It has the characteristics of high efficiency and high quality for welding thick plates. Due to the increased production of thick plate welded structures, NGW technology was popularized in many fields in the early 1980s, such as petrochemical high-pressure machines, power plant boiler thick-walled cylinders, nuclear reactor vessels, turbine shafts, and large steel structures. The ultra-narrow gap welding (ultra-NGW) method further reduces the gap of the weld, which greatly reduces the cross-sectional area of the welding and simplifies the groove processing, and also reduces the cost of welding materials and power consumption. The arc current used in the welding process is less dense and generates less heat input, thus reducing the risk of sheet deformation and residual stress. Compared with traditional welding technology, NGW can reduce the plastic and toughness damage to the heat-affected zone and increase the service life. At the same time, the thick plate adopts a narrow and deep welding groove, which can reduce the stomata defects. The HAZ is narrow and the grains obtained are fine, which improves the welding quality. Considering the small groove angle and small gap, how to ensure side-wall fusion, improve fusion welding formation, and prevent weld cracks is a challenge faced by the ultra-NGW [2].

Zhang et al. [3] developed four technologies including the multi-functional integrated ultra-narrow gap mixed gas arc welding (ultra-NGMAGW) gun, adaptive intelligent tracking of the weld trajectory, efficient gas protection of the high-temperature welding zone, and high stability of the droplet transition, which put ultra-NGMAGW welding technology into production. At present, Zhang et al. have successfully welded some typical steel such as 35Cr2Ni2Mo [4] and 35CrMoV [5] using ultra-NGMAGW, and their research has shown that ultra-NGMAGW can significantly improve the strength of the weld. It is because the ultra-NGMAGW has a lower groove filling area and less heat input than the NGSAW, which further reduces the impact of high temperature on the HAZ during welding and ensures the performance of the welded joint.

A large number of research experiments have shown that the test steel containing AF has satisfactory mechanical properties [6–9]. At the same time, Li et al. [10] found that there were AF with fine grains in the welded joints of the NGW and characterized the mechanical properties, and then found that the strength of the welded joints was higher than that of the bulk material and it had good plasticity. In this paper, low carbon steel Q235A was selected as the welding object, and the ultra-NGMAGW and the NGSAW welding experiments were carried out. We characterized the microstructure and mechanical properties of the welded joints under the two welding methods and obtained the temperature field distribution and thermal cycle curve in the ultra-NGMAGW via numerical simulation, accordingly, and analyzed the mechanism of tissue evolution and strengthening.

2. Materials and Methods

The base material (BM) used in this study is a Q235A low carbon steel with a dimension (length × width × thickness) of 216 × 150 × 18 mm, respectively. The chemical composition of the BM is shown in Table 1, and all concentrations are defined in weight percentage. The microstructure of the BM after corrosion with the 3% nitrate alcohol solution is shown in Figure 1, which is a mixture of pearlite and ferrite. The filler wire used in this study is ER90S-G welding wire with a diameter of 1.2 mm, and the chemical composition of the welding wire is shown in Table 2.

Table 1. Chemical composition of Q235A steel (wt%).

C	Si	Mn	S	P	Nb	Fe
0.14–0.22	0.30	0.3–0.65	0.05	0.04	0.15–0.16	Bal.

The equipment used in this ultra-NGMAGW experiment is a six-axis cross-section of the ultra-narrow gap automatic welding equipment. The schematic diagram of the experimental device is shown in Figure 2. Special ceramic insulation components are installed at the end of the welding gun near the high-temperature area. The welding gun body is made of metal materials with good conductivity and thermal conductivity, and technical measures such as self cooling water circulation and efficient heat dissipation are adopted inside the body to ensure the reliability and stability of the welding process under NG-GMAW conditions.Reduce spatter and improve the droplet transition stability by precisely adjusting the ratio of mixed gases (Ar, CO_2, O_2, and He) using a gas-shielded welding power supply with a high quality cooperative control function. Based on contact sensing, the welding gun is designed to float between states to carry out adaptive weld tracking based on the elastic pressure balance principle. The schematic diagram of the groove after processing is shown in Figure 3. This groove is designed to make the blunt edge of the root of the weld more transparent and not to produce non-fusion of the side wall during welding.

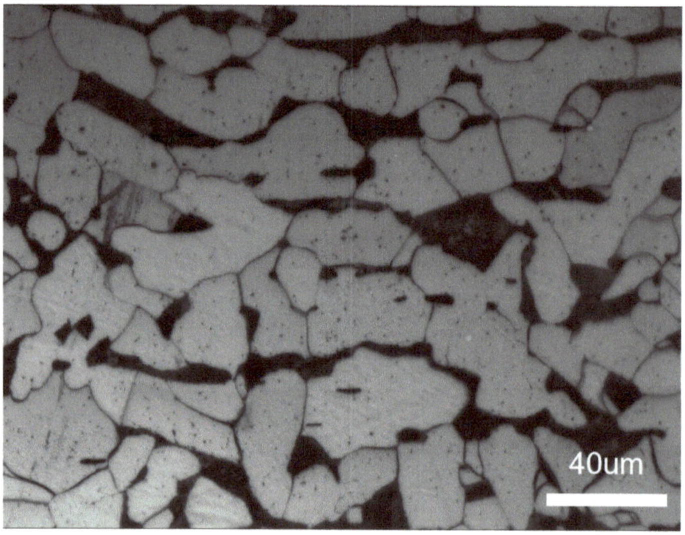

Figure 1. Microstructure of Q235A base metal.

Table 2. Chemical composition of ER90S-G (wt%).

C	Si	Mn	S	Cr	Ni	Mo	V	W	Cu	Fe
0.06	0.40	0.78	0.001	2.68	0.06	0.01	0.21	1.69	0.04	Bal.

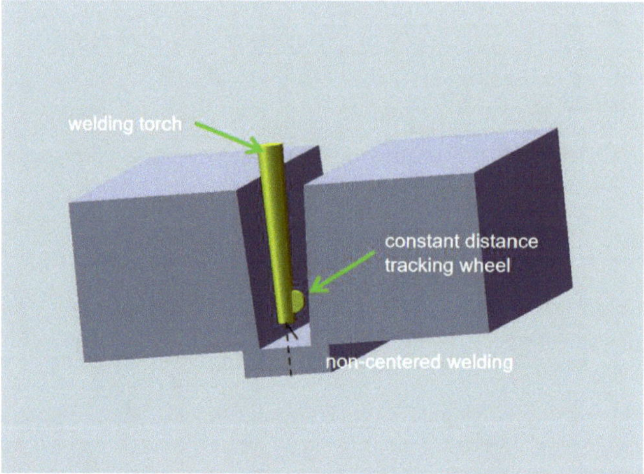

Figure 2. Schematic diagram of the ultra-NGMAGW device.

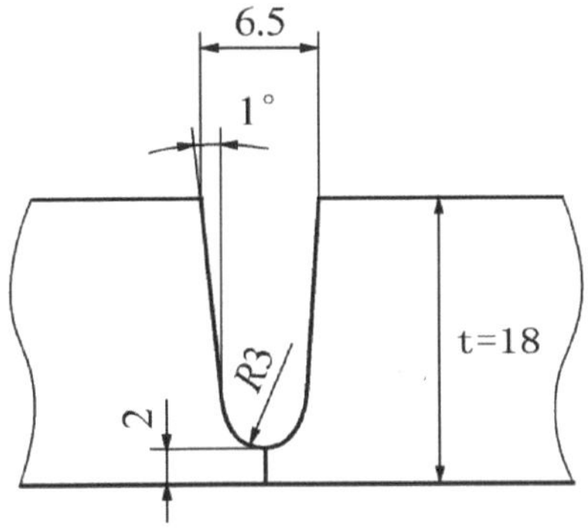

Figure 3. Schematic diagram of the ultra-NGMAGW welding groove.

Two 216 × 75 × 18 mm Q235A thick plates were joined with the above groove, and the area to be welded was polished to remove the oxide and impurities to improve the welding quality. The clamp was used to hold the base material for the ultra-NGMAGW. The detailed parameters of the ultra-NGMAGW are shown in Table 3.

Table 3. Welding process parameters of the ultra-NGMAGW.

Arc Voltage	Welding Current	Welding Speed	Protective Gas	Gas Flow
26.4 V	195 A	4.64 mm/s	80% (Ar) + 20% (CO_2)	20 L/min

After welding, the welded specimens were sectioned transverse to the welding direction and then mounted, ground, polished, and etched with a 3% Nital solution. In this study, the CarlZeissAxioLab.A1 optical microscope (OM) (Carl Zeiss, Oberkochen, Germany) and the TESCANMIRA3LMH mode field-emission scanning electron microscope (SEM) (TESCAN, Brno, Czech Republic) were used to characterize the microstructure of the specimens, and the grain size was measured from the OM image via an area equivalence. Electron back scattering diffraction (EBSD) was also used to characterize the grain size, orientation, and dislocation density in the weld zone. EBSD was performed using a detector field emission SEM with an acceleration voltage of 20 KV and a minimum scanning step size of 20 nm. The specimens were analyzed using HKL channel5 software 5.12.74.0. For measuring the grain size via EBSD, the grain was defined using a misorientation larger than 2° and the average grain size was calculated via an area equivalence.

Microhardness was measured from the weld zone to the base material zone of each sample with a load of 100 g and duration of 10 s using an HXV-1000A microhardness tester (Laizhou Lyric Testing Equipment Co., Ltd., Laizhou, China), and the indentation interval was set to 0.24 mm. All specimens were prepared using standard polishing procedures. The microhardness tester was calibrated with the standard specimens.

For the uniaxial tensile test, 2 mm thick specimens with geometry and dimensions as shown in Figure 4 were cut from the selected area of the weld via electron-discharging machining. All specimens were polished with 1000 mesh sandpaper and then cleaned and dried with anhydrous ethanol to minimize the surface roughness. Tensile tests were conducted at room temperature on a universal testing machine (MTS E45) at a loading speed of 0.5 mm/min. Both specimens were tested at least three times.

In order to study the formation of AF in the ultra-NGMAGW, we use the finite element analysis software Abaqus to simulate the temperature field distribution of Q235A low carbon steel during the ultra-NGMAGW. First, construct the Q235A thick plate model, which is sized to 216 × 150 × 18 mm, and then divide the model into hexahedral elements. The welding model is shown in Figure 5. And, the number of nodes and the number of grids is 21,759 and 19,656, respectively.

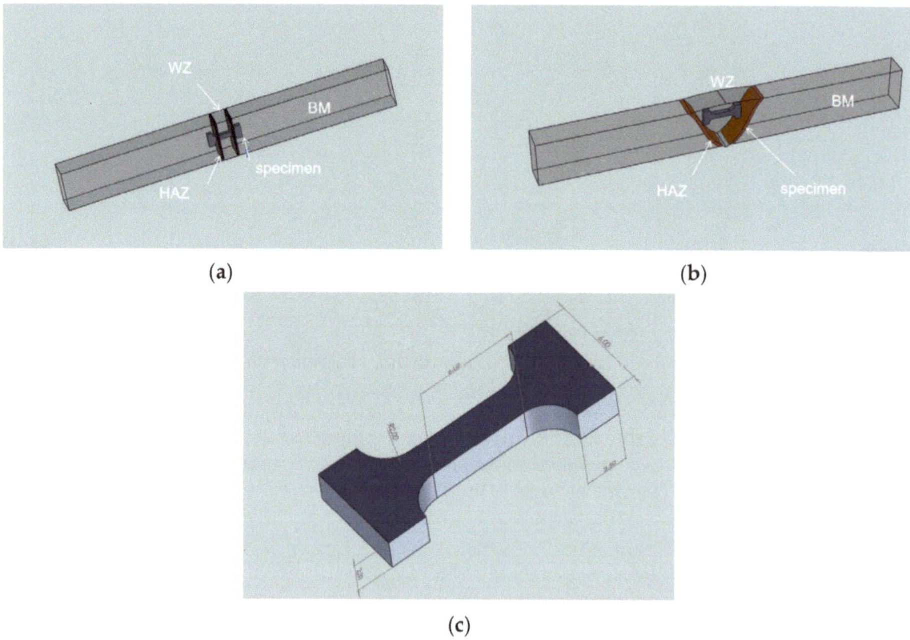

Figure 4. Cut position and size of tensile specimen: (**a**) sampling position of the ultra-NGMAGW; (**b**) sampling position of the NGSAW; and (**c**) sample size (mm).

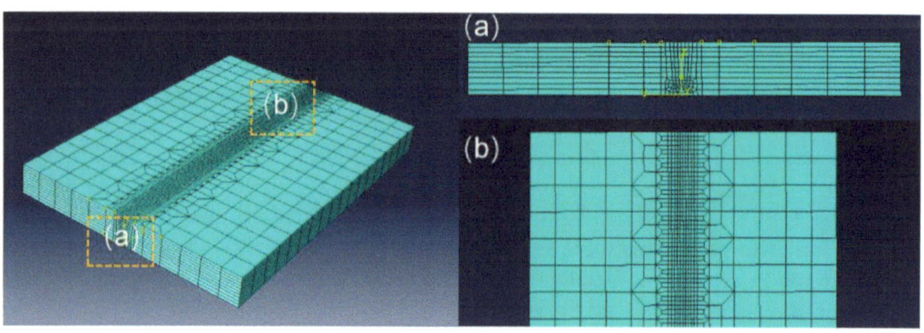

Figure 5. Q235A thick plate joint finite element model. (**a**) weld cross-section; (**b**) weld surface.

According to the heat input characteristics of the ultra-NGMAGW, the double ellipsoid heat source model is selected for simulation. The heat flux distribution equation of the double ellipsoidal heat source model is divided into the front hemisphere and the back hemisphere, as shown in Equation (1).

$$\begin{aligned} q_1(x,y,z,t) &= \frac{6\sqrt{3}f_1Q}{a_1bc\pi\sqrt{\pi}}e^{-3(x+vt)^2/a_1{}^2}e^{-3y^2/b^2}e^{-3z^2/c^2} \\ q_2(x,y,z,t) &= \frac{6\sqrt{3}f_2Q}{a_2bc\pi\sqrt{\pi}}e^{-3(x+vt)^2/a_2{}^2}e^{-3y^2/b^2}e^{-3z^2/c^2}, \end{aligned} \quad (1)$$

where Q = ηP, η is the thermal efficiency, P is the laser power, v is the welding speed, t is the welding time, a_1, a_2, b, c are the geometric parameters of the double ellipsoid heat source model, and f_1 and f_2 are the power distribution coefficients of the front and back semi-ellipsoid, respectively, generally $f_1 + f_2 = 2$. The welding heating process is a nonlinear transient heat conduction process which satisfies Fourier's law, and the temperature field calculation formula is shown in Equation (2).

$$\rho(T)\frac{\partial T}{\partial t} = Q(x,y,z,t) + \frac{\partial\left(\lambda(T)\frac{\partial T}{\partial x}\right)}{\partial x} + \frac{\partial\left(\lambda(T)\frac{\partial T}{\partial y}\right)}{\partial y} + \frac{\partial\left(\lambda(T)\frac{\partial T}{\partial z}\right)}{\partial z} \tag{2}$$

where $Q(x,y,z,t)$ is the internal heat source intensity, T is the temperature, t is the time, ρ is the material density, $\lambda(T)$ is the thermal conductivity of the material, and ∂T is the specific heat capacity of the material. In addition, it is necessary to determine the initial conditions and boundary conditions in the welding process. When $t = 0$, the initial temperature of the weldment is 20 °C. The boundary condition is the heat exchange between the surface of the test plate and the surrounding environment. For the analysis of the welding thermal process, the third type of boundary condition is considered, as shown in Equation (3):

$$\lambda\frac{\partial T}{\partial n} = h\left(T_w - T_f\right) \tag{3}$$

where λ is the thermal conductivity of the object, T_w is the surface temperature of the object, T_f is the temperature of the medium around the object, and h is the heat transfer coefficient between the surface of the object and the surrounding medium. According to the actual welding conditions, the temperature of the medium around the weldment is set at 20 °C. In this paper, the convection heat transfer coefficient is defined as 18×10^{-6} W/(mm^2·°C), and the solid line temperature is 1450 °C. Due consideration is given to the role of thermal radiation, and the radiative rate is defined as 0.8.

The ultra-NGMAGW involves multiple-pass filled bevelings. The upper pass does not have more additional heat input than the lower pass, while the lower pass is subject to additional heat cycling due to the filling of the upper pass. This simulation adopts the death and death unit technology. After the beginning, the elements of the first weld will be awakened and generate heat, and the subsequent elements will be awakened one after another as the heat source moves. After the simulation of the first weld is completed, the calculation of the next weld will be awakened, and the above procedure will be repeated until the simulation of the three welds is completed.

3. Results

3.1. Structure and Morphology of Welded Joints

The cross-section profiles of the welded joints of 18 mm thick Q235A thick plates after the ultra-NGMAGW and the NGSAW are respectively shown in Figure 6. The ultra-NGMAGW specimens showed a "U" shape fusion zone, while the NGSAW specimens showed a "V" shape fusion zone. The Ultra-NGMAGW joint can be clearly distinguished via three weld paths.

The microstructure of the weld zone (WZ) obtained via the ultra-NGMAGW and NG-SAW are presented in Figure 7. Figure 7a,b shows the ultra-NGMAGW WZ microstructure, mainly composed of acicular ferrite (AF), bulk ferrite (BF), and a small fraction of granular bainite (GB). Figure 7c,d shows the microstructure of the NGSAW WZ joint, mainly composed of proeutectoid ferrite (PF) and pearlite (P). However, as shown in Figure 7a,c, it can be found that the weld grain obtained via the ultra-NGMAGW is finer than that obtained via the NGSAW. The mean grain size of the ultra-NGMAGW welded joints is about 6.98 µm, while it is 12.80 µm in the NGSAW welded joints. In addition, Figure 7b shows that the AF grows in all directions around with inclusions or defects as the core [11,12], and in the growth process, due to the different nucleation directions or large growth angles of AF in the adjacent grains, cross-collision occurs to form a locking mechanism, which hinders the continued growth of the grains. In Figure 7d, only P and PF with a coarse grain structure were observed, while AF does not exist. In addition, we can find more dispersed carbides existing in the ultra-NGMAGW joints compared to the NGSAW joints.

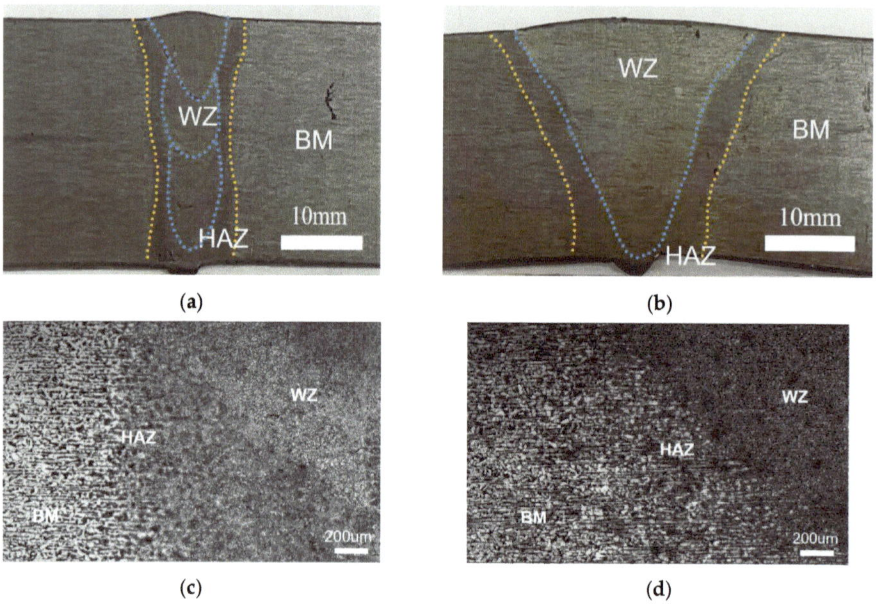

Figure 6. Cross-section profiles and its 50× enlarged view of Q235A welded joints: (**a**) cross-section profiles of Ultra-NGMAGW; (**b**) cross-section profiles of NGSAW; (**c**) enlarged view of Ultra-NGMAGW; and (**d**) enlarged view of NGSAW.

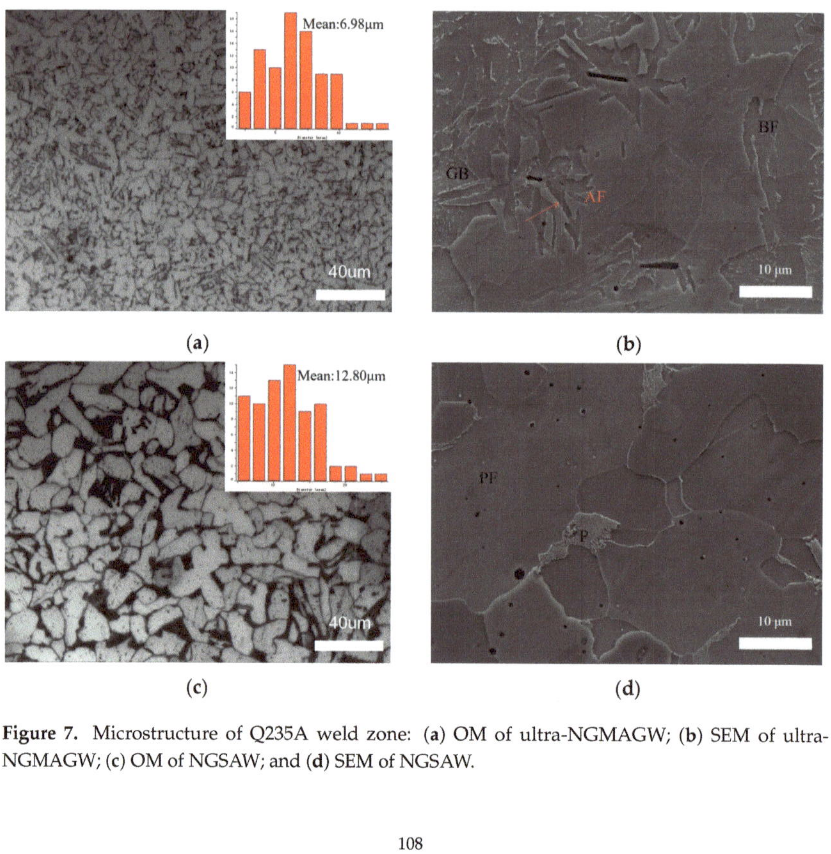

Figure 7. Microstructure of Q235A weld zone: (**a**) OM of ultra-NGMAGW; (**b**) SEM of ultra-NGMAGW; (**c**) OM of NGSAW; and (**d**) SEM of NGSAW.

The microstructure of the heat-affected zone (HAZ) obtained via the ultra-NGMAGW and the NGSAW is shown in Figure 8. Figure 8a,b shows the microstructure of the HAZ of the ultra-NGMAGW sample, mainly composed of GB, BF, and a small amount of AF and P. Figure 8c,d are the microstructures of the HAZ of the NGSAW sample, mainly composed of PF and P. Compared with Figure 8a,c, it can be found that the HAZ grains obtained via the ultra-NGMAGW are finer than those obtained via the NGSAW. The mean grain size of the ultra-NGMAGW HAZ is about 28.35 μm, while it is 33.34 μm in the NGSAW HAZ. This is mainly due to the fact that, as shown in Figure 8b, a small amount of AF and GB are interwoven into the tissues obtained from the ultra-NGMAGW, while coarse grain PF in the tissues are obtained from the NGSAW as shown in Figure 8d, and a large amount of pearlite exists between the ferrites.

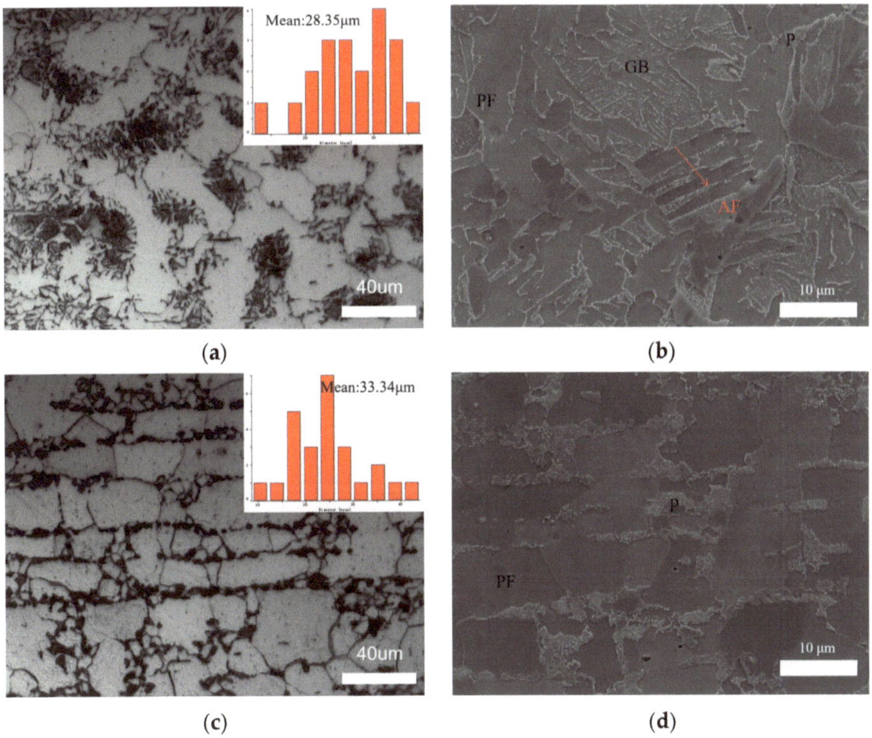

Figure 8. Microstructure of Q235A HAZ: (**a**) OM of ultra-NGMAGW; (**b**) SEM of ultra-NGMAGW; (**c**) OM of NGSAW; and (**d**) SEM of NGSAW.

In summary, it can be found that the microstructure of both the WZ and HAZ obtained via the ultra-NGMAGW is more uniform and finer than that obtained via the NGSAW, which is mainly due to the presence of AF in the joint obtained via the ultra-NGMAGW.

3.2. Mechanical Properties

3.2.1. Microhardness

The Vickers hardness values of the Q235A welded joints were determined after the ultra-NGMAGW and the NGSAW were respectively measured via the HVS-1000A digital microhardness tester (Laizhou Lyric Testing Equipment Co., Ltd., Laizhou, China). The hardness maps of the entire weld for both the ultra-NGMAGW and NGSAW joints are shown in Figure 9a,b, respectively. Due to the hardness variation in different zones, the WZ, HAZ, and BM can be clearly distinguished from the hardness contour map.

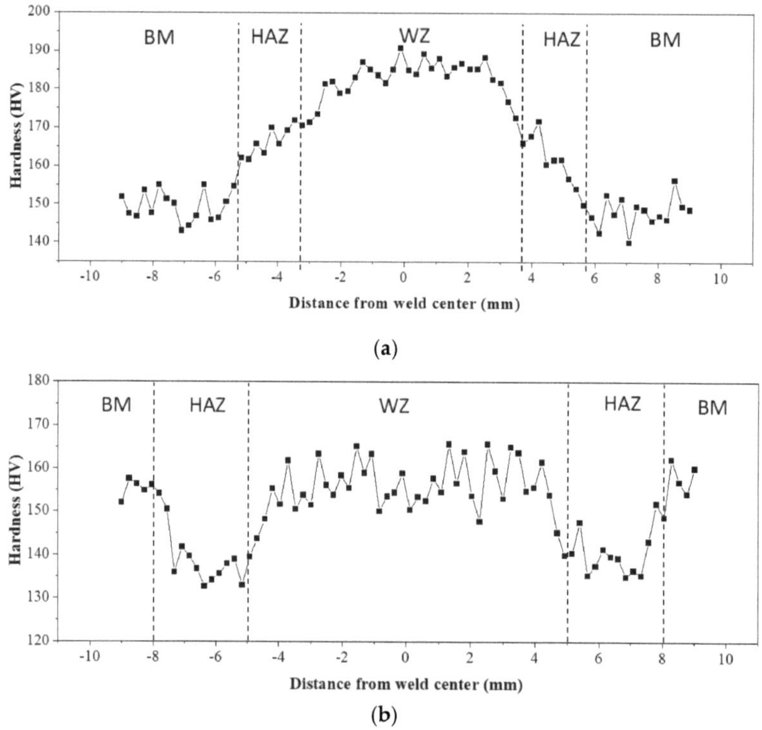

Figure 9. Microhardness measurements on the welded joint (WZ: weld center; HAZ: heat-affected zone; BM: base material): (**a**) ultra-NGMAGW, and (**b**) NGSAW.

The results show that the average hardness of the BM region of both the two specimens is about 155HV0.1, while the average hardness in the WZ and HAZ regions of the ultra-NGMAGW specimen is higher than that of the NGSAW specimen. The average hardness in the WZ for the ultra-NGMAGW sample is about 185HV0.1, and in the HAZ, it is about 165HV0.1. Obviously, the peak hardness is located in the WZ. The NGSAW sample has a hardness of 155HV0.1 in the WZ and a hardness of 140HV0.1 in the HAZ.

By comparing Figure 9a,b, it can be found that there is a certain degree of softening in the joint tissue of the NGSAW; this may be due to the precipitation ferrite, as shown in Figure 8d, induced by the thermal welding cycle in the area of the HAZ and the fusion line as well as the presence of soft ferrite, which leads to the microhardness drop in these regions. There is no such phenomenon in the ultra-NGMAGW. According to Figures 7 and 8, the increase in hardness in the WZ is related to the generation of AF, while the increase in hardness in the HAZ is the result of a large amount of GB structure transformation and a small amount of AF.

3.2.2. Tensile Property

Tensile tests were conducted on the Q235A welded joints of the ultra-NGMAGW and the NGSAW, respectively, and the fractured samples are shown in Figure 10. It can be seen that the ultra-NGMAGW joint failed in the BM well away from the weld, which indicates that the Q235A did not weaken after the ultra-NGMAGW, while the NGSAW joint failed at the weld, which indicates that the NGSAW joint was not as strong as the BM.

Figure 11 shows the representative engineering stress versus the engineering strain curves for the BM, and the ultra-NGMAGW and NGSAW Q235 steel welded joints. The details of the transverse tensile test results are given in Table 4. The yield strength (YS), ultimate tensile strength (UTS), and apparent elongation for the ultra-NGMAGW joints were obtained as 504 MPa, 643 MPa, and 34.5%, respectively, and the NGSAW specimens demonstrated a YS of 327 MPa, a UTS of 487 MPa, and an apparent elongation of 53.2%, respectively. While the YS, UTS, and apparent elongation of the BM were 281 MPa, 464 MPa, and 36.1%, respectively. The Ys of both the two kinds of welded joints are apparently higher than the BM. However, under the premise that the elongation is close to the BM, the Ultra-NGMAGW joint performs more excellent tensile strength, which corresponds to 139% of that of the base metal, and 132% of that obtained via the NGSAW. And, the NGSAW joints performs very high elongation.

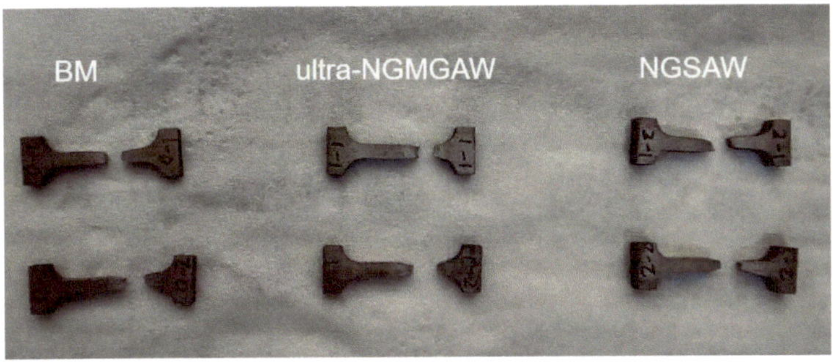

Figure 10. Tensile specimen after fracture.

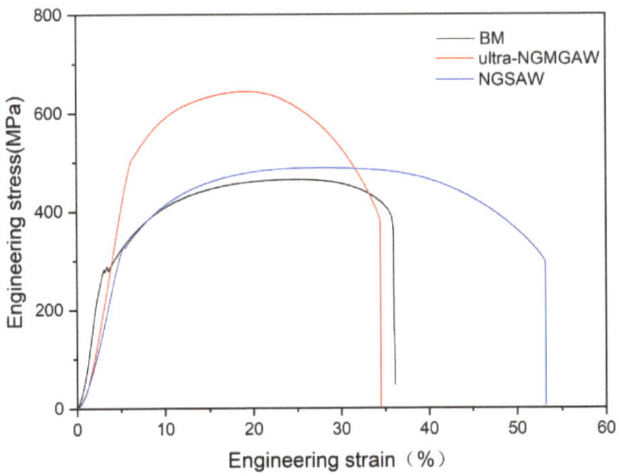

Figure 11. Representative engineering stress versus engineering strain curves for BM, and Ultra-NGMAGW and NGSAW Q235A joints.

Table 4. Tensile properties for the ultra-NGMAGW, NGSAW, and BM.

	Yield Strength (MPa)	Tensile Strength (MPa)	Elongation (%)
BM	281	464	36.1
ultra-NGMAGW	504	643	34.5
NGSAW	327	487	53.2

The fracture surface morphology of the welded joints obtained via SEM is depicted in Figure 12. High magnification observations reveal that all the ductile fracture regions present dimples and microvoids, with some of the dimples containing spherical inclusions, so they all show a ductile fracture mode. While the dimples in the BM are obviously bigger and deeper from Figure 12a, the dimples in the ultra-NGMAGW are finer and more uniform. This also explains that the welded joints obtained via the ultra-NGMAGW own the highest tensile strength.

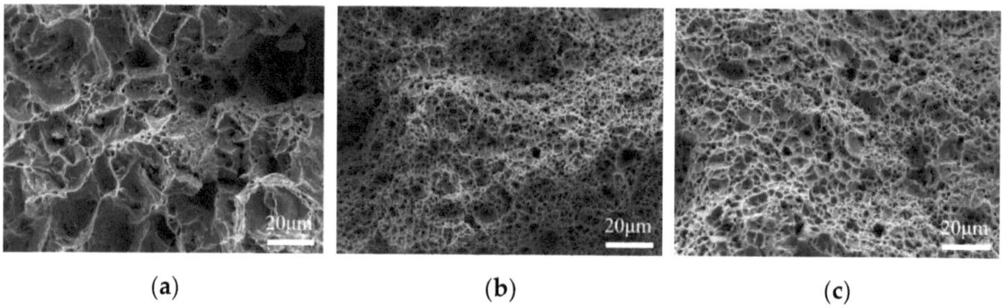

Figure 12. SEM micrographs of the tensile fracture surfaces of (**a**) BM; (**b**) ultra-NGMAGW; and (**c**) NGSAW.

Using the above experimental results, it can be found that the WZ and the HAZ of the welded joint obtained via the ultra-NGMAGW has higher strength and hardness. According to Xiao et al. [13], the microstructure characteristics of AF can improve the strength and toughness of the steel to a certain extent, which is consistent with the observed microstructure in Figures 7 and 8. It can be considered that the strengthening in the ultra-NGMAGW joint comes from AF.

4. Discussion

4.1. Reinforcement Mechanism

In previous analyses, we have known that the excellent mechanical properties of the welded joints obtained via ultra-NGMAGW are mainly due to the formation of AF, which in turn produces grain refinement in the welded joint structure. In order to further analyze the effect of the ultra-NGMAGW welding method on the crystal's characteristics and reveal the relationship between the welding method and the mechanical properties, the specimens after the ultra-NGMAGW and NGSAW were analyzed via EBSD.

Figure 13 shows the WZ microstructure inverse pole figure and corresponding grain size distribution diagram obtained via ultra-NGMAGW and NGSAW, respectively. Figure 13a,b shows the inverse pole of two types of welding microstructures. According to the statistical results of the grain size in Figure 13c,d, it can be found that the average grain size of the WZ structure obtained via the ultra-NGMAGW is about 7 μm, which is basically consistent with the aciculate ferrite size found in the study [14], and the average grain size of the WZ structure obtained via NGSAW is 11 μm. Obviously, the grains obtained via the ultra-NGMAGW is finer.

Figure 14 shows the grain boundary characteristic distribution diagram at the WZ of the two welding methods and the statistical diagram of the misorientation angle of adjacent grain boundaries. The low-angle boundary with adjacent grain misorientation between 2–15° are marked with red lines, and the high-angle boundary with adjacent grain misorientation greater than 15° are marked with black lines. According to the grain boundary characteristic distribution and the grain boundary orientation distribution results, the number of high-angle boundaries in the weld of the specimens after ultra-NGMAGW is significantly higher than that after NGSAW. In previous work [15], high-angle boundaries can inhabit crack propagation and change its direction, so the strength of the ultra-NGMAGW welded joints are obviously higher than the NGSAW.

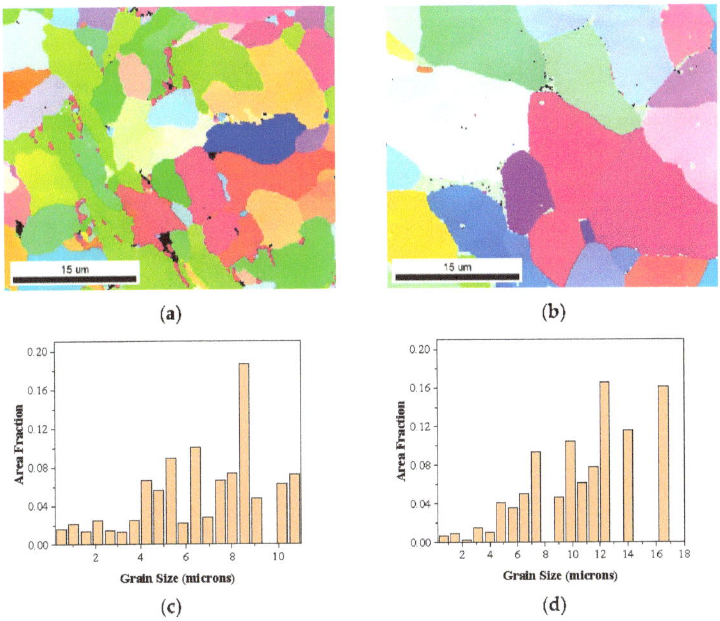

Figure 13. Inverse pole figure and grain size distribution diagram at the WZ: (**a**) inverse pole figure of the ultra-NGMAGW joint; (**b**) inverse pole figure of the NGSAW joint; (**c**) grain size distribution diagram of the ultra-NGMAGW joint; and (**d**) grain size distribution diagram of the NGSAW joint.

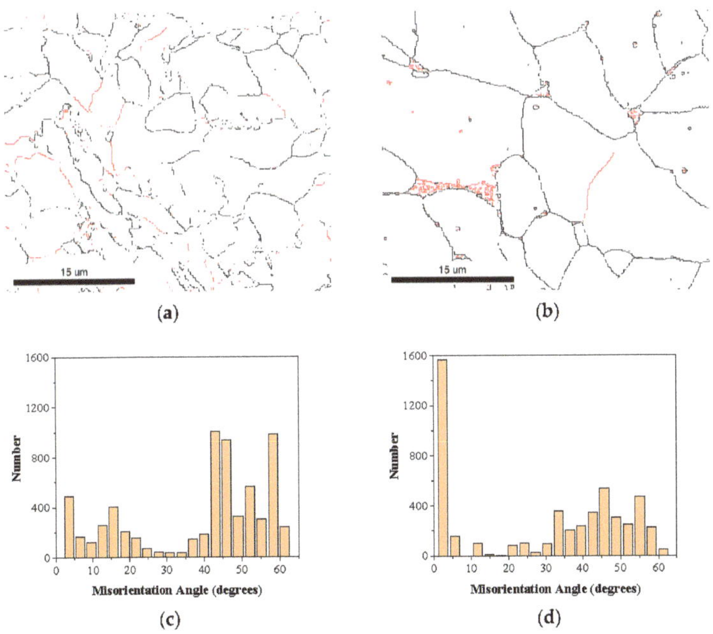

Figure 14. Adjacent grain boundary misorientation distribution map and angle statistical map at the WZ: (**a**) adjacent grain boundary misorientation distribution map of ultra-NGMAGW joint; (**b**) adjacent grain boundary misorientation distribution map of the NGSAW; (**c**) angle statistical map of the ultra-NGMAGW; and (**d**) angle statistical map of the NGSAW.

The distribution of grain boundary misorientation in the weld microstructure can reflect the magnitude of dislocation density inside the weld microstructure. Figure 15 shows the kernel average misorientation diagram of two types of welding microstructures, and the kernel average misorientation (KAM) is defined as the misorientation between each data point and its neighboring points.

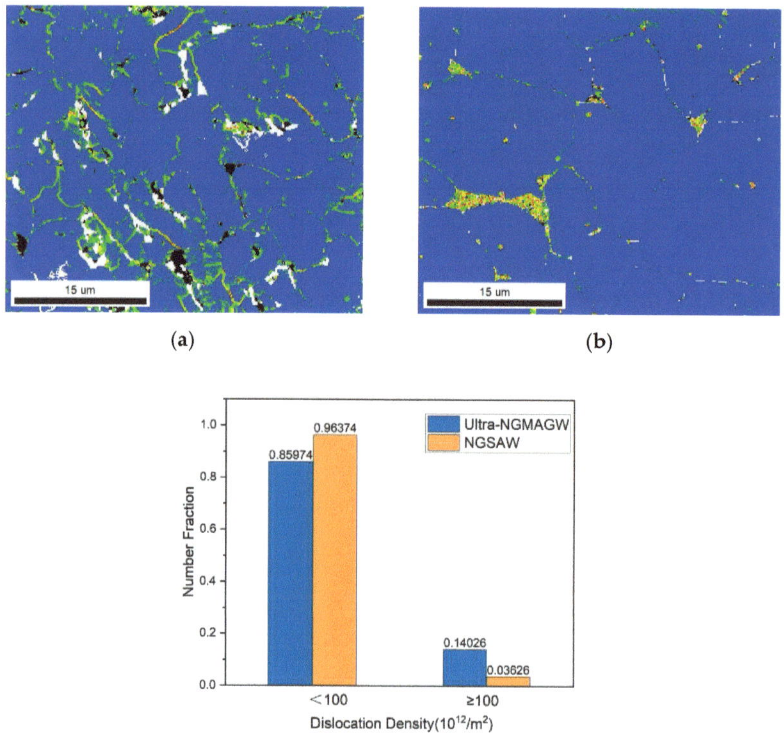

(c)

Figure 15. The distribution diagram of the KAM and statistical diagram of dislocation density in the WZ: (**a**) the KAM of the ultra-NGMAGW; (**b**) the KAM of the NGSAW; and (**c**) the statistical diagram of dislocation density in the ultra-NGMAGW and NGSAW.

As we know from Figure 7, the ultra-NGMGAW WZ joint is mainly composed of AF and BF, while the NGSAW WZ joint is mainly composed of BF and P. There are only a few high-density dislocation regions at the junction of BF. As can be seen from the local misorientation results in Figure 15c, the proportion of weld dislocation density above 100 obtained via the ultra-NGMGAW can reach nearly 15%, while the proportion of weld dislocation density above 100 obtained via the NGSAW is only close to 5%. The dislocation density in the region of the AF distribution is significantly higher than that of the BF distribution. AF has a high density of dislocation inside, so when the welded joint is deformed via external force, it is easy to emit a dislocation at the tip of the crack, forming a packet structure of the dislocation and the timely passivation of the crack so that the crack growth is blocked.

In summary, the welded joints obtained via the ultra-NGMAGW own higher strength than that of the NGSAW, which is mainly due to the AF obtained during the ultra-NGMAGW. As we know, there are a large number of dispersed carbide phases in the ultra-NGMAGW playing a dispersion strengthening role, which has also been studied in previous work [16]. The grain size obtained of the ultra-NGMAGW weld joints is smaller than that of the NGSAW. According to the Hall–Petch relationship [17], the welded joints obtained via the ultra-NGMAGW own higher strength, which reflects the effect of the fine grain strengthening of AF corresponding to Figures 10 and 14. In addition, the smaller the grain size, the greater the number of high-angle boundaries. At the same time, due to the smaller grain size in the Ultra-NGMAGW, it increases the mean volume fraction of grain boundaries, which may act as barriers and provide obstacles to the propagation of brittle cracks [15,18], so grain boundary strengthening is also reflected in the ultra-NGMAGW. And, in general, the high angle grain boundaries effectively resist the crack propagation and change its direction [19]. The number of high-angle boundaries in the weld of the specimens after the ultra-NGMAGW is significantly higher than that after the NGSAW, so the effect of grain boundary strengthening is more obvious in the ultra-NGMAGW joints. Additionally, as we know, when the dislocation density in steel is relatively large, it will hinder the slip of the dislocation and improve the strength of the steel. It is precisely because of the presence of AF that the weld obtained via the ultra-NGMAGW can improve the mechanical properties of the weld under the co-action of fine grain strengthening, grain boundary strengthening, and dislocation strengthening.

4.2. Numerical Simulation and Microstructure Evolution

The mechanical properties of the welded metal are closely related to their microstructures, which are dependent on the chemical composition of the material and the thermal history (cycles) due to the welding processes [20,21].

According to Figures 7 and 8, we found that in the ultra-NGMAGW, the pearlite and ferrite in the BM undergo structural transformation, where GB and a small amount of AF appear in the HAZ, while the pearlite changes to a large amount of AF and a few GB in the WZ. AF is the main factor affecting the mechanical properties of the welded joints.

As seen in Figure 16a,b, the EDS results of the welded joints obtained via the ultra-NGMAGW and the NGSAW are shown. The content of Si and Mn in the specimens of the ultra-NGMAGW is significantly higher than that of the NGSAW. The alloying elements in steel promote the nucleation and refinement of AF [22]. That is to say, the higher composition of Si and Mn in the welded joints of the ultra-NGMAGW promotes the formation of AF, which in turn optimizes the mechanical properties of the ultra-NGMAGW joints.

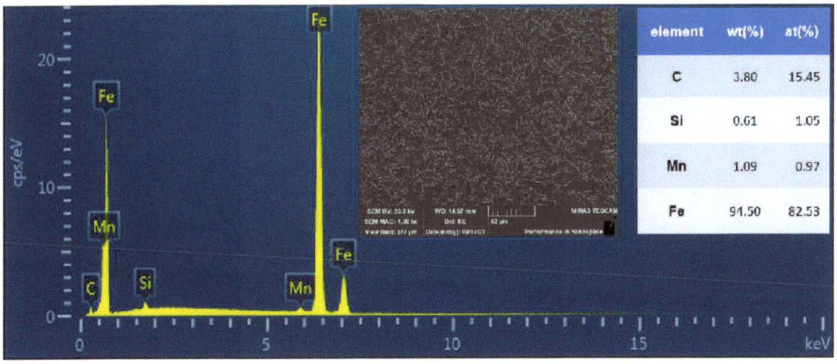

(a)

Figure 16. *Cont.*

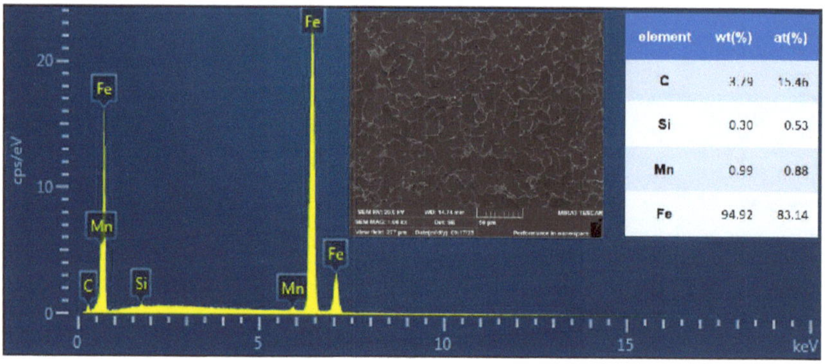

(**b**)

Figure 16. (**a**) EDS spectrum for the ultra-NGMAGW welded joints, and (**b**) EDS spectrum for the NGSAW welded joints.

The weld temperature distribution results of the Q235A ultra-NGMAGW joints obtained via finite element simulation are shown in Figure 17. The simulated temperature field distribution shape diagram of the weld is in good agreement with the positions of the weld zone and the heat-affected zone in the actual welding process, indicating that the simulation is effective.

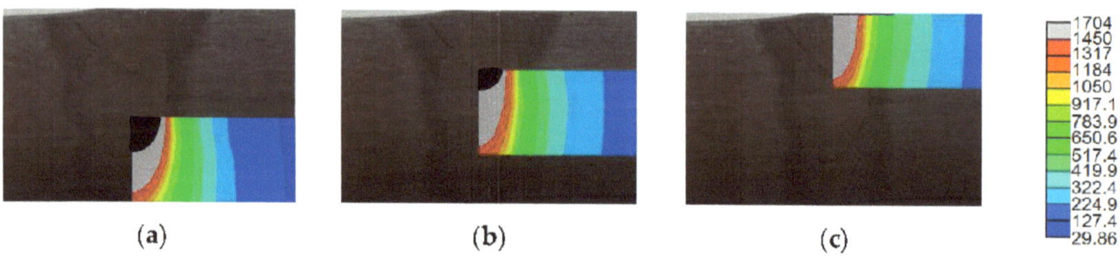

(**a**) (**b**) (**c**)

Figure 17. Temperature distribution of welded joint: (**a**) first weld; (**b**) second weld; and (**c**) third weld.

The temperature at the center of the three passes is selected as the temperature of the WZ and the thermal cycle curve is obtained, which is shown in Figure 18.

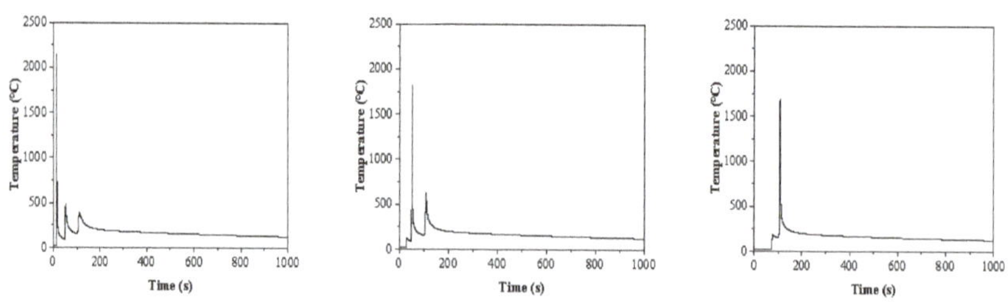

(**a**) (**b**) (**c**)

Figure 18. Welding thermal simulation of thermal cycle curve in the weld zone: (**a**) the first weld; (**b**) second weld; and (**c**) third weld.

Figure 18a shows that during the first welding, the temperature of the weld zone of the weld bead rapidly rose to 2140.3 °C, and then rapidly decreased to 152.5 °C. After that, the second welding was performed in Figure 18b. This time, the temperature of the weld zone of the weld bead rose to 1827.6 °C, and the first weld was heated to 488 °C, and then quickly decreased to 152.2 °C. For the third welding in Figure 13c, the temperature of the weld in this pass was increased to 1684.4 °C, and the first two welds were heated to a maximum of 387.0 °C and 639.2 °C, respectively, and then rapidly cooled to 115.2 °C after 1000 s.

It can be seen from the thermal cycle curve of the weld zone that due to the small heat input during the welding process, the temperature of the weld zone will drop rapidly after rising to the maximum height and will not stay on the Ac3 line for too long. In addition, the subsequent pass welding will not heat the previous pass temperature above the Ac1 line, so the influence of the subsequent pass on the microstructure evolution of the previous pass can be ignored.

According to studies [23,24], various types of medium temperature transition structures will be generated during the continuous cooling of low-carbon steel. Granular bainite structures are mainly obtained at a low cooling rate, and lath bainite structures are mainly obtained at a high cooling rate. During continuous cooling, AF can be formed in the higher transition temperature range.

Combined with the weld thermal cycle curve in Figure 18, and the SEM micrographs in Figure 7b, it can be concluded that austenite grains will transform into different ferrite structures during the process of the temperature drop after the ultra-NGMAGW. First of all, a very small part of austenite is transformed into pearlite during the rapid cooling process, and then transforms to bulk ferrite as the temperature drops. As the temperature drops to about 600 °C, more AF nucleates at the inclusion [25], and granular bainite appears after continuous cooling.

While in the NGSAW, combined with the SEM micrographs in Figure 7d and due to the high heat input and low cooling rate of the NGSAW, pearlite and bulk ferrite with coarse grains will be formed, and these coarse grains will reduce the strength and toughness of the weld.

5. Conclusions

(1) AF exists in the microstructure of the ultra-NGMAGW welded Q235A joints, and the grain size of AF is smaller than that of pearlite and bulk ferrite in the NGSAW welded joints.

(2) The welded joints obtained via the ultra-NGMAGW have better mechanical properties, and the hardness and strength are higher than those obtained via the NGSAW.

(3) The main reason for the formation of AF is that the heat input in the welding process is small, the cooling rate is fast, and the high temperature residence time is short.

(4) AF with a small grain size, high-angel boundary, and high dislocation density exists in the joints obtained via the ultra-NGMAGW, which results in excellent deformation resistance and mechanical properties under the co-action of fine grain strengthening, grain boundary strengthening, and dislocation strengthening mechanism.

Author Contributions: Conceptualization, S.W. and W.X.; methodology, F.Z.; software, S.W. and L.G.; validation, S.W. and L.G.; formal analysis, S.W.; investigation, S.W.; resources, W.X.; data curation, S.W.; writing—original draft preparation, S.W.; writing—review and editing, S.W. All authors have read and agreed to the published version of the manuscript.

Funding: This research received no external funding.

Institutional Review Board Statement: Not applicable.

Informed Consent Statement: Not applicable.

Data Availability Statement: Data available on request due to restrictions eg privacy or ethical. The data presented in this study are available on request from the corresponding author.

Conflicts of Interest: The authors declare no conflict of interest.

References

1. Li, X.D.; Ma, X.P.; Subramanian, S.V.; Misra, R.D.K.; Shang, C.J. Structure–Property–Fracture Mechanism Correlation in Heat-Affected Zone of X100 Ferrite—Bainite Pipeline Steel. *Metall. Mater. Trans.E.* **2015**, *2*, 1. [CrossRef]
2. Sun, Q.J.; Hu, H.F.; Yuan, X.; Feng, J.C. Research Status and Development Trend of Narrow-gap TIG Welding. *Adv. Des. Technol.* **2011**, *308*, 1170–1176. [CrossRef]
3. Zhang, F.; Guo, J.; Zhang, G.; Wang, Y. Narrow Gap/Ultra Narrow Gap Welding Technology (IV) Ultra Narrow Gap MAG/MIG Welding Technology. *Electr. Weld. Mach.* **2017**, *47*, 25–29.
4. Guo, J.; Zhang, F.; Pang, H.; Zhang, G. Analysis of microstructure and properties of ultra narrow gap MAG welded joints between 34Cr2Ni2Mo and 35CrMoV steel thick plates. *Electr. Weld. Mach.* **2018**, *48*, 51–54+60.
5. Guo, J.; Wan, W.; Zhang, F.; Zhang, G. Microstructure analysis of ultra-narrow gap MAG welding joints in 35CrMo quenched and tempered steel. *Electr. Weld. Mach.* **2017**, *47*, 24–27.
6. Zhao, M.-C.; Ke, Y.; Shan, Y.-Y. Comparison on Strength and Toughness Behaviors of Pipeline Steels with AF and Ultrafine Ferrite. *Mater. Lett.* **2003**, *57*, 1496–1500. [CrossRef]
7. Byun, J.-S.; Shim, J.-H.; Suh, J.-Y.; Oh, Y.-J.; Cho, Y.W.; Shim, J.-D.; Lee, D.N. Inoculated AF Microstructure and Mechanical Properties. *Mater. Sci. En. A Struct. Mater. Prop. Microstruct. Process.* **2001**, *319*, 326–331. [CrossRef]
8. Zhao, M.; Xiao, F.; Shan, Y.; Li, Y.; Yang, K. Microstructural Characteristic and Toughening of an Ultralow Carbon AF Pipeline Steel. *Chin Shu Hsüeh Pao* **2002**, *38*, 283–287.
9. Lan, L.; Qiu, C.; Zhao, D. The Formation of AF and Mechanical Properties Behaviors in a Low Welding Crack Susceptibility Steel. *Adv. Mater. Res.* **2011**, *163–167*, 337–341.
10. Zhao, L.; Zhao, X. Microstructure and Mechanical Properties of Weld in High Strength Low Alloy Steel by Ultra-narrow Gap Welding. *Adv. Mater. Res.* **2011**, *154–155*, 1792–1795. [CrossRef]
11. Song, M.; Song, B.; Zhang, S.; Xue, Z.; Yang, Z.; Xu, R. Role of Lanthanum Addition on AF Transformation in C-Mn Steel. *ISIJ Int.* **2017**, *57*, 1261–1267. [CrossRef]
12. Cui, Z.M.; Zhu, L.G.; Li, Y.L.; Zhang, Q.J.; Yan, C.L.; Mo, W.L. Relationship between Crystal Structure of Inclusions and Formation of AFs. *J. Iron Steel Res. Int.* **2016**, *23*, 586–592. [CrossRef]
13. Xiao, F.R.; Zhao, M.C.; Shan, Y.Y.; Liao, B.; Yang, K. Processing of Ultralow Carbon Pipline Steels with AF. *J. Mater. Sci. Technol.* **2004**, *20*, 779–781.
14. Wu, K.M. Three-dimensional Analysis of AF in a Low-carbon Steel Containing Titanium. *Scr. Mater.* **2006**, *54*, 569–574. [CrossRef]
15. Ghosh, A.; Ghosh, M. Tensile and Impact Behaviour of Thermo Mechanically Treated and Micro-alloyed Medium Carbon Steel Bar. *Constr. Build. Mater.* **2018**, *192*, 657–670. [CrossRef]
16. Ghosh, A.; Ghosh, M.; Gudimetla, K.; Kalsar, R.; Kestens, L.A.I.; Kondaveeti, C.S.; Pugazhendhi, B.S.; Ravisankar, B. Development of Ultrafine Grained Al–Zn–Mg–Cu Alloy by Equal Channel Angular Pressing: Microstructure, Texture and Mechanical Properties. *Arch. Civ. Mech. Eng.* **2020**, *20*, 7. [CrossRef]
17. Chokshi, A.H.; Rosen, A.; Karch, J.; Gleiter, H. On the validity of the Hall-Petch relationship in nanocrystalline materials. *Scr. Met.* **1989**, *23*, 1679–1684. [CrossRef]
18. Ghosh, A.; Ghosh, M.; Shankar, G. On the Role of Precipitates in Controlling Microstructure and Mechanical Properties of Ag and Sn Added 7075 Alloys during Artificial Ageing. *Mater. Sci. Eng. A Struct. Mater. Prop. Microstruct. Process.* **2018**, *738*, 399–411. [CrossRef]
19. Chen, J.; Tang, S.; Liu, Z.Y.; Wang, G.D. Microstructural characteristics with various cooling paths and the mechanism of embrittlement and toughening in low-carbon high performance bridge steel. *Mater. Sci. Eng. A* **2013**, *559*, 241–249. [CrossRef]
20. Chen, Y.; Wang, Y.-Y.; Gianetto, J. Thermal and microstructure simulation of high strength pipeline girth welds. In Proceedings of the Eighteenth International Offshore and Polar Engineering Conference, Vancouver, BC, Canada, 6–11 July 2008.
21. Lampman, S.R. *Weld Integrity and Performance, ASM International Technology& Engineering*; Material Park: Novelty, OH, USA, 1997.
22. Vezzù, S.; Scappin, M.; Boaretto, D.; Timelli, G. On the Effect of Slight Variations of Si, Mn, and Ti on Inclusions Properties, Microstructure, and Mechanical Properties of YS460 C-Mn Steel Welds. *Metallogr. Microstruct. Anal.* **2019**, *8*, 292–306. [CrossRef]
23. Shang, C.; Hu, L.; Yang, S.; Wang, X.; Zhao, Y.; He, X. Forming and Controlling of the AF in Low Carbon Steel. *Chin Shu Hsüeh Pao* **2005**, *41*, 471–476.
24. Zhao, M.; Shan, Y.; Qu, J.; Xiao, F.; Zhong, Y.; Yang, K. AF Formation in a Pipeline Steel with Thermo-mechanical Control Process. *Chin Shu Hsüeh Pao* **2001**, *37*, 820–824.
25. Kim, Y.M.; Lee, H.; Kim, N.J. Transformation Behavior and Microstructural Characteristics of AF in Linepipe Steels. *Mater. Sci. Eng. A Struct. Mater. Prop. Microstruct. Process.* **2008**, *478*, 361–370. [CrossRef]

materials

MDPI

Article

Hot-Cracking Mechanism of Laser Welding of Aluminum Alloy 6061 in Lap Joint Configuration

Km Rakhi [1], Seunggu Kang [1] and Joonghan Shin [1,2,3,*]

1 Department of Future Convergence Engineering, Kongju National University, 1223-24 Cheonandaero, Seobuk-gu, Cheonan 31080, Republic of Korea; rakhi@smail.kongju.ac.kr (K.R.); topkun97@gmail.com (S.K.)
2 Department of Mechanical and Automotive Engineering, Kongju National University, 1223-24 Cheonandaero, Seobuk-gu, Cheonan 31080, Republic of Korea
3 Global Institute of Manufacturing Technology (GITECH), Kongju National University, 1223-24 Cheonandaero, Seobuk-gu, Cheonan 31080, Republic of Korea
* Correspondence: jhshin@kongju.ac.kr; Tel.: +82-41-521-9254; Fax: +82-41-555-9123

Abstract: Laser welding, known for its distinctive advantages, has become significantly valuable in the automotive industry. However, in this context, the frequent occurrence of hot cracking necessitates further investigation into this phenomenon. This research aims to understand the hot-cracking mechanism in aluminum alloy (AA) 6061, welded using a laser beam in a lap joint setup. We used an array of material characterization methods to study the effects of processing parameters on the cracking susceptibility and to elucidate the hot-cracking mechanism. A laser power of 2000 W generated large hot cracks crossing the whole weld zone for all welding speed conditions. Our findings suggest that using a heat input of 30 J/mm significantly mitigates the likelihood of hot cracking. Furthermore, we observed that the concentrations of the alloying elements in the cracked region markedly surpassed the tolerable limits of some elements (silicon: 2.3 times, chromium: 8.1 times, and iron: 2.7 times, on average) in AA6061. The hot-cracking mechanism shows that the crack initiates from the weld root at the interface between the two welded plates and then extends along the columnar dendrite growth direction. Once the crack reaches the central region of the fusion zone, it veers upward, following the cooling direction in this area. Our comprehensive investigation indicates that the onset and propagation of hot cracks are influenced by a combination of factors, such as stress, strain, and the concentration of alloying elements within the intergranular region.

Keywords: laser welding; aluminum alloy 6061 (AA6061); lap joint; cracking susceptibility; hot-cracking mechanism

1. Introduction

The aluminum 6061 alloy is favored in a variety of industries, including the automotive, marine, and aerospace industries, owing to its exceptional mechanical properties and superior corrosion resistance compared to other aluminum alloy series [1–3]. Numerous fusion-based welding techniques are employed to join aluminum alloys, among which laser welding is a promising method [4–6]. However, laser welding of these alloys often results in defects such as voids and hot cracks [7–9]. Hot cracking, in particular, is a critical defect likely to occur during the laser welding of aluminum alloys [10,11]. The chemical composition of these alloys increases their vulnerability to hot cracking [12]. Specifically, the morphology of the weld zone considerably influences the formation of hot cracks.

The susceptibility to hot cracking and its underlying mechanisms have been thoroughly studied by many researchers. Rappaz et al. [13] created a numerical model by considering various microstructural and thermal parameters such as the grain morphology, size, and orientation as well as the temperature field and the macroscopic strain rate. This model offers insights into the genesis of hot cracks. Kou et al. [14] demonstrated that hot cracking is more likely to occur within a partially melted zone owing to an increased vulnerability to the tensile stresses and strains generated during solidification. This is attributed to the formation of residual stresses during the phase transition from liquid to solid, which can lead to cracking if the yield strength of the material is surpassed. Liu et al. [15] introduced a thermomechanical model by considering the movement of impurities within a material, influencing its mechanical and thermal properties, and employed strain theory to assess the cracking susceptibility of a material. Wang et al. [16] formulated a finite element model that simulated aluminum alloy behavior during laser welding, emphasizing strain localization in the mushy transitional zone. This model predicts the likelihood of cracking during welding by evaluating the local deformation of the material. Various other hot-cracking models and criteria have been explored and analyzed in the referenced studies [17–21].

Despite numerous studies on hot cracking during the laser welding of aluminum alloys, there is a need for additional research. Most of these studies have concentrated on the butt joint configuration. In contrast, lap joint laser welding—common when joining structural components such as automotive chassis or body panels—remains less explored. This technique, employed with alloys such as AA6061, facilitates lightweight construction and a high strength, stiffness, and corrosion resistance. Thus, it is critical to comprehend the mechanisms of hot cracking within this context, and further research is needed to devise effective prevention strategies.

In this work, we aimed to study the hot-cracking mechanism for the laser lap joint welding of AA6061. A series of experiments were designed and conducted to generate both sound (without hot cracking) and defective (with hot cracking) welds. The effects of the processing conditions on the cracking susceptibility were investigated, and various material characterization methods were used to elucidate the hot-cracking mechanism. It is expected that the proposed hot-cracking mechanism in this study contributes to enhancing the understanding of the hot-cracking phenomenon in the laser welding of AA6061 under a lap joint configuration.

2. Material and Methods

2.1. Material

The experimental samples were rectangular plates of AA6061, each 2 mm thick and artificially aged to the T6 condition (AA6061-T6). The dimensions of each plate were 70 mm in length and 40 mm in width. Prior to welding, the specimen surfaces were meticulously cleaned with acetone to prevent any surface contamination issues. Table 1 provides the elemental composition of AA6061.

Table 1. Chemical composition of AA6061.

Material	Al	Si	Fe	Cu	Mn	Mg	Cr	Zn	Ti	Others
wt. %	97.34	0.71	0.5	0.24	0.12	0.9	0.18	0.08	0.04	0.01
at. %	97.78	0.68	0.24	0.11	0.05	1.0	0.08	0.03	0.03	0.01

2.2. Experimental Procedure

This study focuses on the laser lap welding of an aluminum alloy (AA6061) using a single-mode continuous-wave ytterbium fiber laser system (Raycus, RFL-C2000). This system delivers a maximum power output of 2000 W at a wavelength of 1080 nm. The laser beam, transmitted through an optical fiber, was focused onto the surface of the specimen using a focusing lens with a focal length of 200 mm. The diameter of the beam spot on the workpiece surface was 100 μm. Full penetration of the laser beam into the top plate is required during the laser lap welding process to ensure the complete fusion of the base metal. To achieve this, we identified suitable process conditions via bead-on-plate experiments to guarantee the complete penetration of the laser beam into the AA6061 plate. Table 2 presents the optimal process conditions. To maintain uniformity and limit distortion and residual stress variations, the plates were secured during the welding experiment. Figure 1 depicts a schematic of the lap joint setup and the clamping conditions.

Table 2. Laser welding process conditions.

Exp. No.	Laser Power (W)	Welding Speed (mm/s)	Heat Input (J/mm)	Power Density ($\times10^5$ W/mm^2)	Interaction Time * ($\times10^{-3}$ s)
A1		47	30		2.13
A2	1400	35	40	1.78	2.86
A3		28	50		3.57
B1		53	30		1.89
B2	1600	40	40	2.03	2.50
B3		32	50		3.13
C1		60	30		1.67
C2	1800	45	40	2.29	2.22
C3		36	50		2.78
D1		67	30		1.49
D2	2000	50	40	2.54	2.00
D3		40	50		2.50

* Interaction time: beam spot size/welding speed.

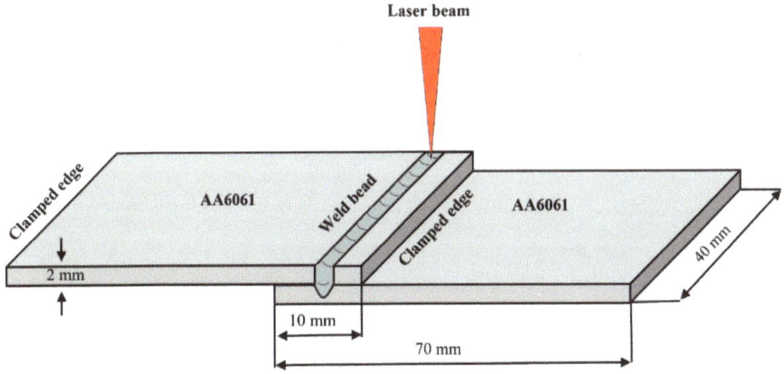

Figure 1. Schematic of lap joint weld configuration.

2.3. Sample Preparation for Weld Morphology and Microstructure Analysis

Following the welding experiment, the welded samples were sectioned perpendicular to the welding direction and secured using a fixation clip, in conjunction with a mixture of epoxy resin and a hardener, a method known as cold mounting. The secured samples were then subjected to grinding with sandpaper and polishing with a diamond suspension up to 1 μm utilizing a polishing cloth. To facilitate the observation of the weld morphology and conduct a microstructural analysis, the polished surfaces of the samples were chemically etched using Keller's reagent, composed of 5 mL of NO_3, 3 mL of HCL, 2 mL of HF, and 20 mL of H_2O. The surface morphologies of the weld bead and weld cross-section were scrutinized using optical microscopy (OM; Dino Lite, Torrance, CA, USA, AM4013MZT) and scanning electron microscopy (SEM; TESCAN, Warrendale, PA, USA, VEGA3). To measure the top weld bead width, five points along the welding direction were selected (each 2.5 mm apart on the image), and the average value from these measurements determined the bead width. An elemental composition analysis of the weld fusion zone (FZ) was performed using energy-dispersive X-ray spectroscopy (EDS; BRUKER, Billerica, MA, USA, flash6130). The microstructure, grain morphology, and grain size were examined using electron backscatter diffraction (EBSD; HIKARI SUPER, Nagoya, Japan, SU5000), with measurements acquired under a 20 kV electron and a step size of 2 μm.

3. Results and Discussion

3.1. Analysis of Weld Morphology and Hot Cracking

Figure 2 presents optical micrographs of the top weld beads from various weld joints, exhibiting varying widths under distinct heat inputs. Specifically, the width tended to expand with increasing laser power, but it contracted with increasing welding speed. This relationship among weld bead width, laser power, and welding speed is likely due to heat input variations influenced by parameter adjustments. Upon the OM examination of the weld joint, centerline hot cracking was found to occur at a relatively high heat input (50 J/mm). Centerline hot cracking was the most pronounced in samples A3, C3, D1, D2, and D3, while no visible cracks were identified on the weld beads of the remaining samples. For samples D1 and D2, hot cracking primarily stemmed from the relatively high power density (2.54×10^5 W/mm^2), given that the heat input was comparatively low. Conversely, for samples A3, C3, and D3, hot cracking was primarily due to a high heat input (50 J/mm), which created a higher thermal gradient in the material, leading to increased residual stresses and strains and resulting in hot cracking. The sample D3, welded with the highest power density (2.54×10^5 W/mm^2), showed the largest crack width (~160 µm); however, the crack width decreased to ~60 µm when the lowest power density (A3: 1.78×10^5 W/mm^2) was applied. Overall, the susceptibility to hot cracking was found to decrease as the heat input was reduced. Apart from the heat input, the interaction time (the duration of laser beam exposure to the material) can also impact hot cracking; a longer interaction time implies more energy absorption by the material, which escalates the risk of hot cracking. Based on the above analysis and the data presented in Figure 2, we concluded that the welding process parameters for the samples with a laser power of 1600 W were not prone to centerline hot cracking, as no such cracking was observed in samples B1, B2, or B3.

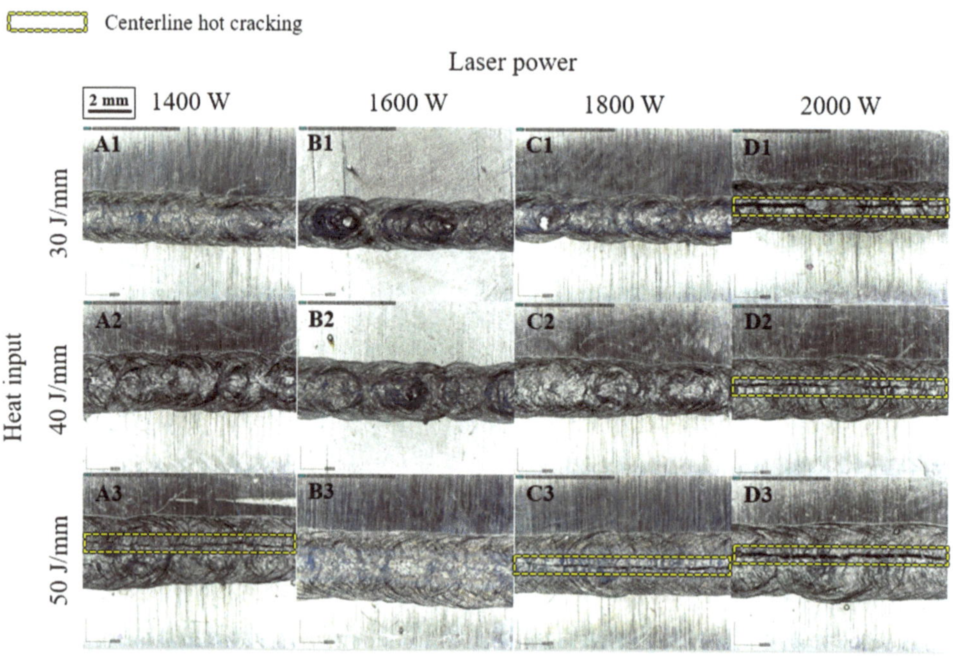

Figure 2. Top surface images of the weld zone.

Figure 3 presents SEM-generated cross-sectional images of the welds, showing fully formed welds in AA6061. An examination of these cross-sections revealed the formation of cracks and voids within the welds. These findings suggest that the penetration depth escalates with increased heat input and laser power, while it diminishes with increasing welding speed. Notably, the cracks were found to originate from the fusion boundary at the interface between the upper and lower plates, propagating towards the FZ center. This occurrence is likely due to the highest thermal stresses being associated with solidification shrinkage near the fusion line, and these stresses can be relayed towards the weld pool center. Upon reaching the center of the FZ, the cracks proceeded to propagate upward along the centerline of the weld. Hot cracking was most conspicuous in the samples with a high heat input (50 J/mm: A3, B3, C3, and D3) and a high power density (2.54×10^5 W/mm^2: D1 and D2). In these cases, the crack length was in the range of ~1.1 to ~2.3 mm. Mild cracking was observed in the samples with a moderate heat input (40 J/mm: A2, B2, and C2), whereas no visible cracks were detected at a low heat input (30 J/mm: A1, B1, and C1). In the case of mild cracking, the measured crack length was in the range of ~130 to ~270 μm. Overall, the susceptibility to hot cracking declined as the heat input decreased. Figure 4 additionally illustrates the impact of the power density and interaction time on hot cracking, demonstrating the relationship between hot cracking and the process parameters. As Figure 4 depicts, the combination of a high power density and an extended interaction time was more susceptible to hot cracking. As this combined value diminished, the susceptibility to hot cracking reduced correspondingly. A high power density implies that a high energy is concentrated on a smaller area for a shorter duration, leading to higher temperatures and faster heating and solidification rates. The rapid heating and cooling rates linked with a high power density generate significant thermal gradients within the metal, resulting in high residual stresses in the solidifying metal that foster hot cracking. An extended interaction time also leads to a high temperature and thermal stress concentration, which in turn promotes hot cracking.

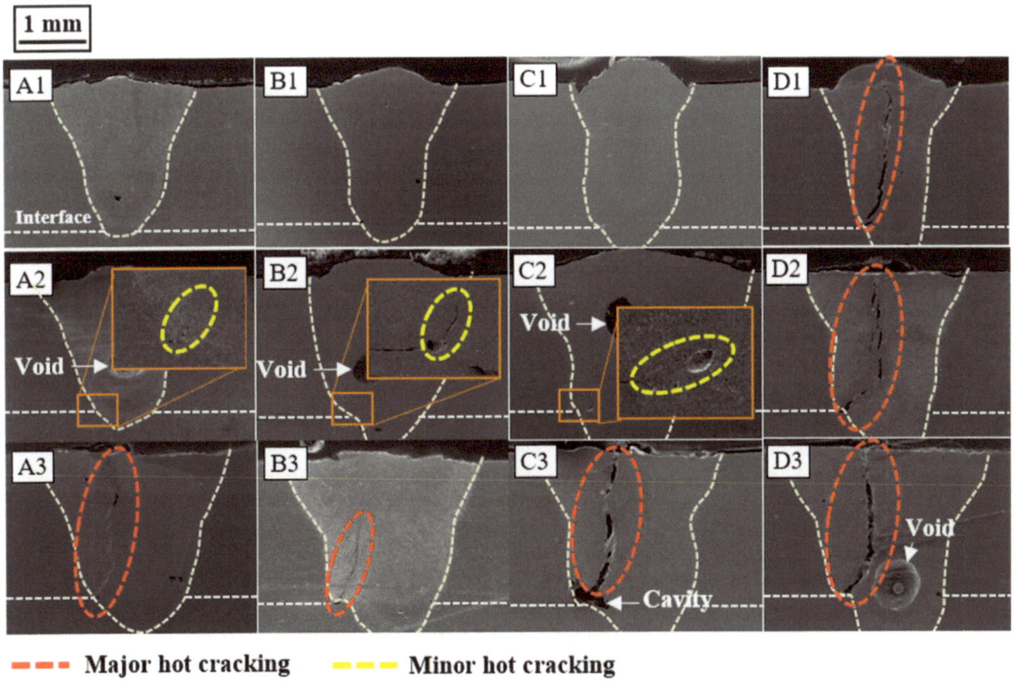

Figure 3. Scanning electron microscopy (SEM) cross-section images of the weld zone.

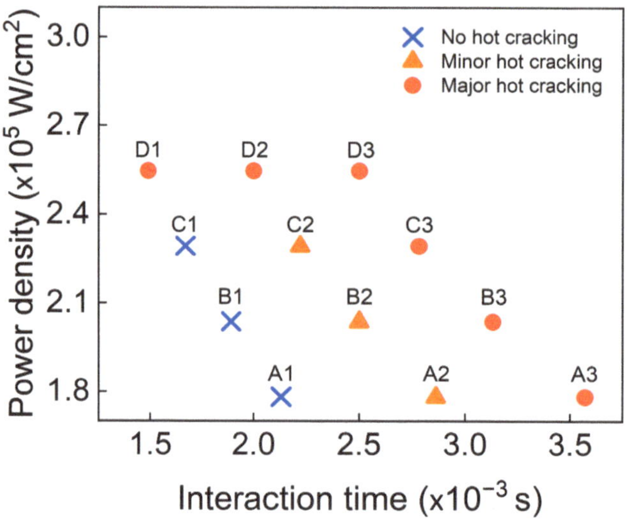

Figure 4. Effect of power density and interaction time on hot cracking.

3.2. Analysis of Microstructure by EBSD

An EBSD analysis was performed to examine the microstructural features of the welds. Figure 5 displays the inverse pole figure (IPF) maps for the transverse cross-section of the weld zone for samples A1, A3, B1, and B3. As presented in Figure 5, the microstructure of the weld zone varied significantly from that of the base metal. Specifically, the grains were finer for a heat input of 30 J/mm compared to 50 J/mm. This observation was ascribed to the higher welding speed associated with a lower heat input and a slower welding speed for a higher heat input. When the welding speed was high, it resulted in an accelerated cooling rate, supporting faster cooling and the formation of smaller grains. Furthermore, a lower heat input implied less heat being applied to the base metal, preventing it from reaching elevated temperatures and further facilitating the formation of finer grains. In the weld zone, columnar grains were predominantly situated at the boundary of the weld, whereas equiaxed grains were largely distributed at the center of the weld. This arrangement resulted from the solidification process of the weld, with columnar grains initially growing from the weld boundary when the weld boundary temperature was at its lowest. These grains persistently grew towards the weld center until the new nucleation of equiaxed grains occurred in the remaining central melt pool. Hence, the solidification process determined the grain distribution. Moreover, Figure 5 indicates that samples A1 and B1 (with a heat input of 30 J/mm) did not present hot cracking, while samples A3 and B3 (with a heat input of 50 J/mm) displayed hot cracking. This phenomenon can be elucidated by the differences in the grain growth mechanisms triggered by varying welding conditions. Solidification, contributing to the formation of larger grains, was implicated in the growth of hot cracks. During solidification, large grains possessed larger intergranular regions. Consequently, they harbored fewer grain boundaries, which created structurally weak areas. The stress concentration was elevated in these intergranular regions, making them act as a path for crack growth. As depicted in Figure 5, hot cracks were apparent in the intergranular regions. After hot cracking, the cracks tracked the columnar grain boundaries towards the center, ultimately propagating in an upward direction.

Figure 6 depicts the fraction of the grain area as a function of the grain diameter. The average grain size for the samples under a low heat input condition (30 J/mm) was 91 μm for A1 and 96 μm for B1, whereas for the samples under a high heat input condition (50 J/mm), it was 93 μm for A3 and 126 μm for B3. Based on the aforementioned analysis, it can be deduced that the grain size expands with increasing heat input. Consequently, the samples subjected to a high heat input and exhibiting large grain sizes experienced significant hot cracking.

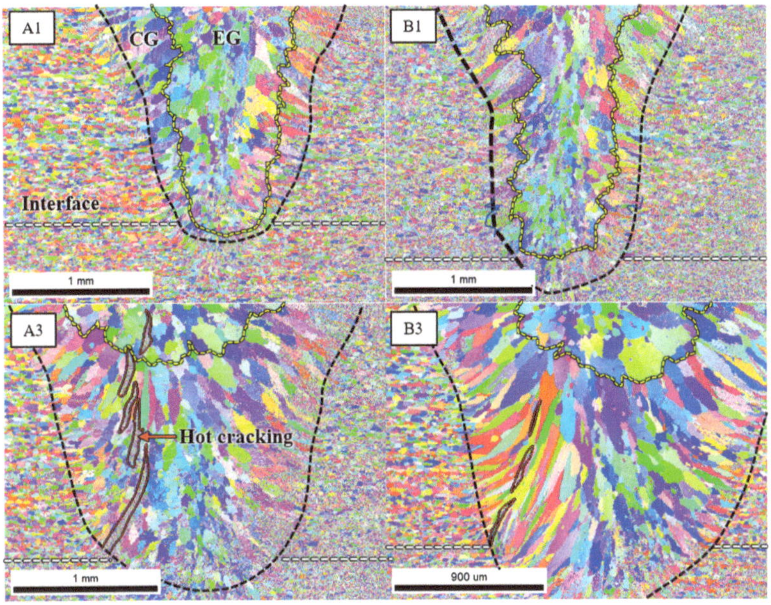

CG: Columnar grains EG: Equiaxed grains

Figure 5. Electron backscatter diffraction (EBSD) inverse pole figure (IPF) maps for transversal cross-section of the weld.

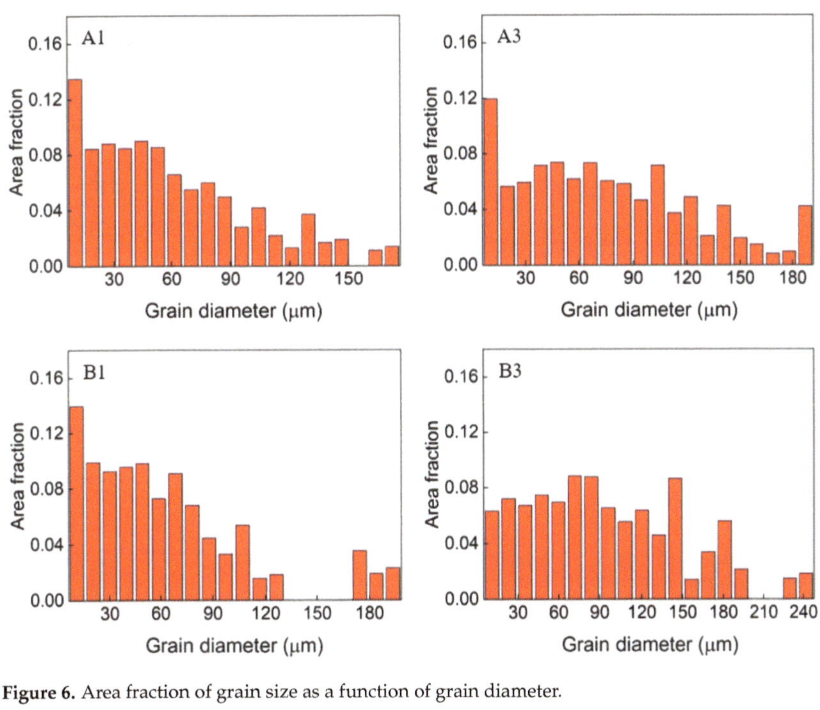

Figure 6. Area fraction of grain size as a function of grain diameter.

3.3. EDS Analysis of Chemical Composition

Using the EDS area and point mapping, the concentrations of the alloying elements in the selected samples were measured. According to the EDS analysis, an instrumental error for the EDS analysis was usually less than 0.1%. Therefore, in this study, we neglected the instrumental error in displaying the results of the EDS analysis. Figure 7 displays the results of the EDS area mapping for samples D1 and D3, both of which exhibited significant hot cracking due to the use of the highest power density (2.54×10^5 W/mm^2). The data show a notably high concentration of silicon (Si) in both samples. For sample D1, the atomic percentages (at. %) of Si were 2.16%, 1.80%, and 1.92% at the initial, middle, and end stages of hot-crack formation, respectively. Similarly, for D3, the Si concentrations were measured as 1.38%, 2.06%, and 1.60% at the same stages. Additionally, the quantities of chromium (Cr) and iron (Fe) were markedly higher in sample D3 than in D1, likely due to a higher heat input during welding. A high heat input implies a slow welding speed (i.e., low solidification rate), allowing a longer period for the alloying elements to segregate in the weld pool. Due to their lower diffusivity in Al, these elements move more slowly with the solidification front compared to Al and become trapped in the interdendritic region. As the weld solidifies, these alloying elements are gradually forced toward the center of the pool as the solidification front moves from the fusion boundary to the center. Conversely, under conditions of a low heat input (i.e., high welding speed), the alloying elements have a comparatively shorter time to segregate throughout the weld, leading to lower concentrations in the samples subjected to a lower heat input compared to those with a higher heat input. However, the copper (Cu) concentration did not show a significant variation. Interestingly, Figure 7 indicates no detectable levels of magnesium (Mg) in samples D1 and D3, likely due to the high vaporization of Mg, which has a relatively low boiling point (1091 °C). This is especially pronounced near the cracks, as they typically form along the centerline, the region with the highest temperature in the weld zone.

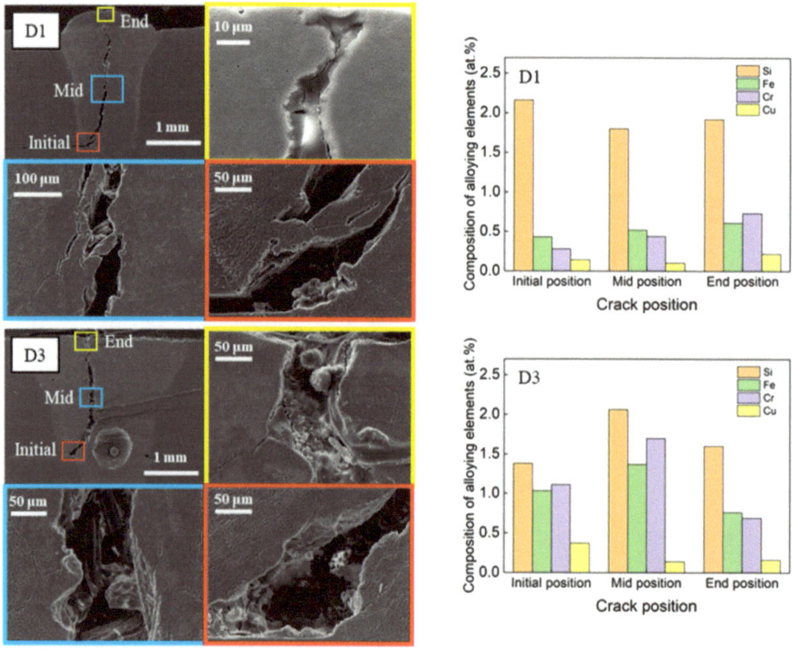

Figure 7. Results of energy-dispersive X-ray spectroscopy (EDS) area mapping at different crack positions for samples D1 and D3.

Figure 8 presents the EDS point-mapping results for samples D1 and D3 in the regions without hot cracking. The analysis showed that the concentrations of the major alloying elements Si, Fe, Cr, and Cu were lower in regions without cracks than those near cracks (as seen in Figure 7), indicating the pivotal role of the chemical composition in hot-crack formation. Although traces of Mg were detected in spectrum S3 for sample D1 and in spectra S1 and S3 for sample D3, the amount of Mg present was minimal. This uneven distribution of Mg in the weld zone can be attributed to several factors, including the vaporization of Mg due to its low boiling point and its segregation or migration near grain boundaries during solidification. As such, a non-uniform distribution of Mg within the weld zone is expected.

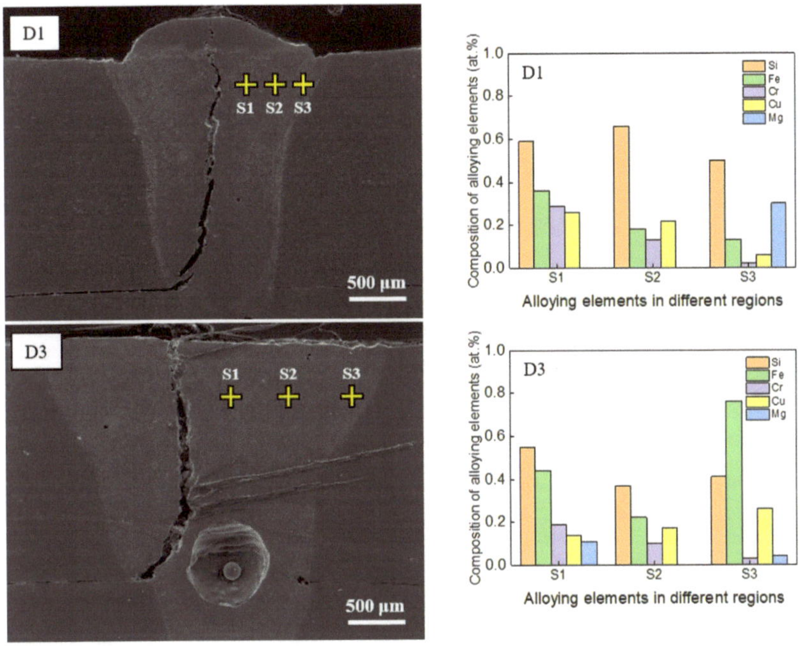

Figure 8. Results of energy-dispersive X-ray spectroscopy (EDS) point mapping in different locations without cracks for samples D1 and D3.

Figure 9 illustrates the results of the EDS point-mapping analysis for the samples that had a low heat input (30 J/mm) and were devoid of hot cracking. The analysis indicated that, for sample A1, the atomic percentages (at. %) of major alloying elements such as Si, Fe, and Cr, excluding Cu and Mg, were less than those of Si, Fe, and Cr in sample D1 (refer to Figure 7, crack region). The diminished concentrations of Cu and Mg in sample D1 were likely because of their vaporization caused by the high temperature in the weld zone, given that sample D1 had the highest laser power (2000 W). This explanation is consistent with the findings for sample B1, where the at. % of Si, Fe, and Cr was also lower than that in sample D1, excluding Cu and Mg. As shown in Figure 9, a considerable Mg concentration was found in the weld zone of sample B1 without hot cracking. According to Wang et al. [22], a high Mg concentration induced a low hot-cracking susceptibility during the directed energy deposition of an Al-Zn-Mg-Cu alloy. Thus, it is considered that the high concentration of Mg contributed to a sound weld formation in the case of sample B1.

As per the American Society for Testing and Materials standards, the permissible thresholds for alloying elements during the welding of AA6061, which minimize the propensity for hot cracking, are outlined in Table 3 [23].

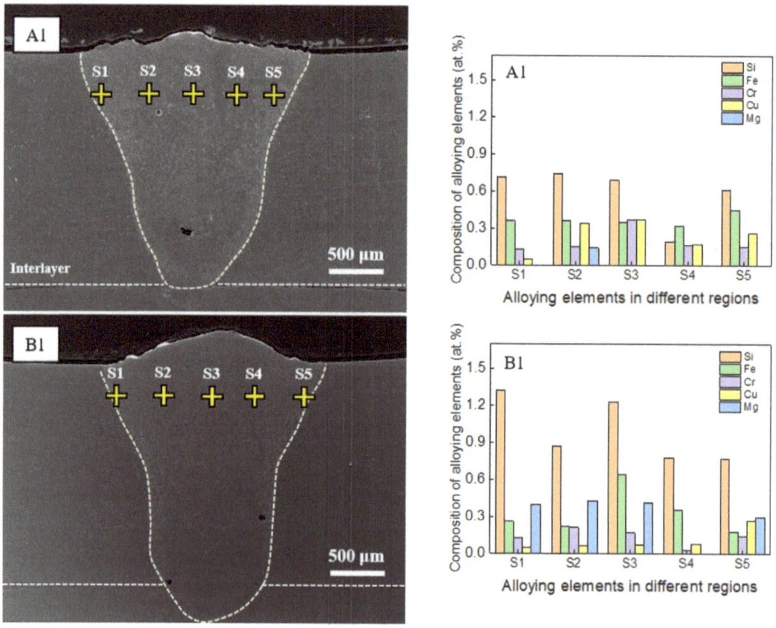

Figure 9. Results of elemental point mapping by energy-dispersive X-ray spectroscopy (EDS) for samples A1 and B1.

Table 3. Acceptable limit for alloying elements in AA6061.

Alloying Element	wt. %	at. %
Si	0.4–0.8	0.38–0.77
Fe	0.7	0.33
Cu	0.15–0.4	0.06–0.17
Mn	0.15	0.07
Mg	0.8–1.2	0.89–1.33
Cr	0.04–0.35	0.02–0.18
Zn	0.25	0.10
Ti	0.15	0.08
Al	97.16	97.59
others	0.2	0.48

Based on the analysis discussed in this section, the heightened vulnerability to hot cracking during the laser welding of AA6061 in a lap joint configuration is linked to high atomic concentrations of Si (ranging from 1.38% to 2.16%), Cr (ranging from 0.76% to 2.16%), and Fe (ranging from 0.69% to 1.11%). The concentrations of these elements in the region prone to cracking significantly exceeded the permissible limits outlined in Table 3, which prescribes general percentage ranges for these alloying elements to minimize the occurrence of hot cracking during welding. Notably, these limits fluctuate depending on the joint configuration and specific welding process parameters. The EDS analysis results suggest that surpassing the composition limits of Si, Cr, and Fe by 2.3, 8.1, and 2.7 times on average, respectively, can increase the risk of hot cracking during the laser welding of AA6061 in a lap joint configuration. Consequently, meticulous attention and control over the levels of these elements during the welding process are strongly recommended to mitigate the likelihood of hot cracking.

3.4. Mechanism of Hot Cracking

The mechanism of hot cracking in the laser welding of AA6061 is primarily determined by various factors such as the properties of the material, the welding parameters, and the joint design. In this study, we propose a mechanism to explain the hot-cracking phenomena. This mechanism, discussing the initiation and growth stages of hot cracking, is detailed in two separate subsections.

3.4.1. Crack Initiation Mechanism

Figure 5 depicts the microstructure of the transverse cross-section of the WZ, composed of columnar and equiaxed grains—a typical structure frequently observed in Al alloys. This is attributed to the high-speed welding process, which results in rapid heat transfer to the base metal. The microstructure of the weld, as illustrated in Figure 5, features columnar grains near the fusion line. In this study, hot cracking was initiated from the root of the weld at the faying interface between the two plates, with the interface serving as a natural pre-crack [24]. This implies that a microcrack was already present at the welding onset.

Figure 10 illustrates a schematic of the crack initiation during the weld pool solidification at various stages. The lap joint configuration with a liquid melt pool is presented in Figure 10a. The pre-crack zone is indicated by a red circle in the diagram. The cooling process, which includes solidification, can be categorized into three stages (see Figure 10b–d).

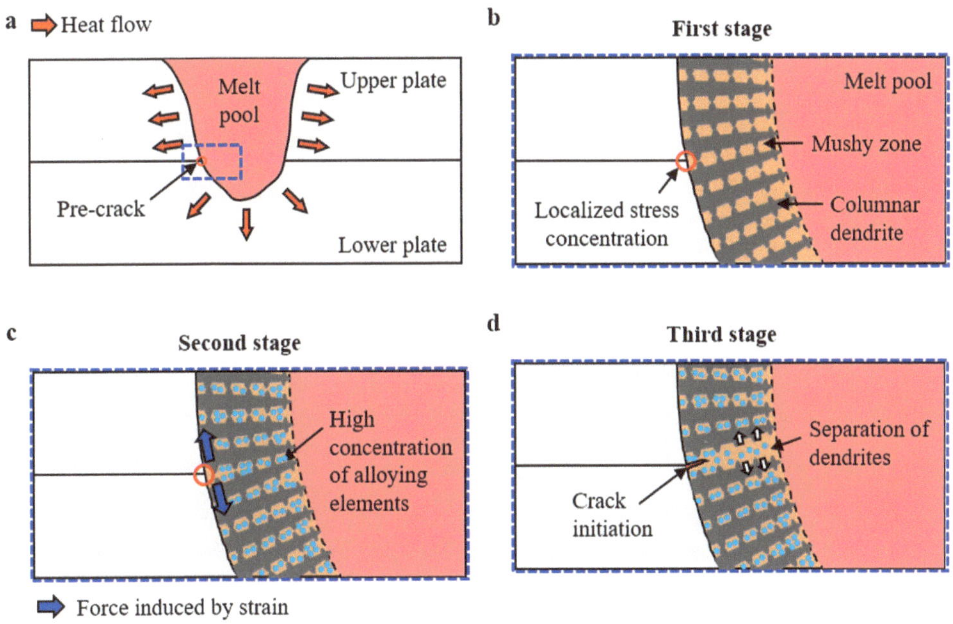

Figure 10. Schematic illustration of crack initiation in the weld zone: (**a**) lap joint configuration with a melt pool; (**b**) localized stress concentration at the interface; (**c**) strain induced between two dendrites due to highly coherent network; (**d**) crack initiation due to separation of dendrites.

In the initial stage, columnar grains primarily form, owing to the temperature gradient and the conditions of solidification experienced by the materials during the process. A higher temperature gradient and slower solidification rate near the fusion boundary encourage the growth of columnar grains. As the heat of the laser beam causes the materials to melt, a melt pool is created. When this pool solidifies, it generates tensile stress due to the shrinkage of the FZ. This stress tends to concentrate at the weld root of the faying interface between the two welding specimens. This concentration occurs because of a sudden geometry change at the weld root, disrupting the stress flow through the material and causing the weld root of the faying interface to act as a stress raiser. Consequently, the stress at the faying interface becomes notably higher than in the surrounding areas, leading to a concentration of stress.

During the solidification of the melt pool, the molten metal transitions from a liquid to a solid phase. As the temperature decreases, columnar dendritic structures start forming at the fusion boundary, extending into the remaining liquid (Figure 10b). Between these columnar dendrites, a mushy zone exists, where both the liquid and solid phases coexist. The width of the mushy zone depends on the freezing range of the alloy [25]. As Al alloys are wide-freezing-range alloys, they possess a relatively broad mushy zone [26]. This expansive mushy zone allows the dendrites of the Al alloys to grow uniformly, forming a coherent network, as shown in Figure 10b. The term coherency of the dendritic network refers to the uniform interconnection of the dendrites.

In the following stage, as illustrated in Figure 10c, the highly coherent network induces strain between two columnar dendrites due to their competing growth for the same space. As a result, the dendrites exert pressure on each other, creating a tendency to separate. As previously discussed, the faying interface at the fusion boundary functions as a pre-crack, leading to stress concentration at the weld root. Moreover, strain occurs between the two dendrites, rendering the weld root at the faying interface susceptible to hot cracking (Figure 10c). Concurrently, the alloying elements (Si, Fe, Cr, and Cu) are trapped in the interdendritic region. Given their low solubility and diffusivity in Al, these alloying elements do not readily blend with Al and segregate in the mushy zone. This segregation behavior results in the trapping of these elements between the dendrites during the rapid solidification of Al, leading to a high concentration of these elements in the remaining mushy zone.

Figure 10d illustrates the final stage of crack initiation. Owing to the strain generated between dendrites, these structures exert pressure on each other, as previously mentioned. The segregation of alloying elements within the mushy zone weakens the cohesive force between neighboring dendrites, facilitating their separation. Furthermore, the stress concentration at the weld root of the faying interface provides a conducive environment for the dendritic separation. As a result, crack initiation commences from the weld root of the faying interface, as depicted in Figure 10d.

3.4.2. Crack Growth Mechanism

Figure 11 presents a schematic illustration of crack growth. Once intergranular hot cracking commences through the solidification of the dendritic structure, hot cracks start to propagate. At this stage, the alloying-element-rich liquid phase in the mushy zone weakens the cohesive force between neighboring dendrites, thus enabling their separation. This leads to inter-dendritic separation, as discussed in Section 3.4.1. When the dendrites separate from each other, the spacing between them increases, promoting the further growth of the intergranular hot cracks (see Figure 11a). This intergranular region of the columnar dendrites provides a path for hot-crack propagation. Consequently, crack propagation tends to align with the growth direction of the columnar dendrites, as illustrated in Figure 11b. When the crack reaches the central region of the FZ, it begins to propagate upward. Solidification generally occurs in the direction opposite to cooling. During cooling, heat dissipates from the top surface (the hottest part of the FZ) to the fusion boundary (the coldest part of the FZ). In the central area of the fusion zone, the direction of cooling is towards the bottom of the fusion zone. As a result, solidification and crack propagation in the central area of the FZ proceed upward. This crack propagation persists until the driving forces (such as a high stress, a high strain, or a high concentration of alloying elements in the intergranular region) for crack growth diminish.

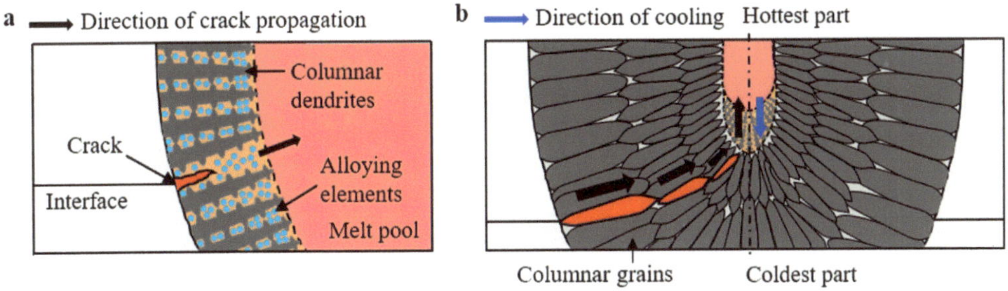

Figure 11. Schematic illustration of crack growth in the weld zone: (**a**) growth of hot crack due to interdendritic separation; (**b**) crack propagation in the FZ.

4. Conclusions

In this study, we examined the laser welding of AA6061 plates in a lap joint configuration. We explored the influence of welding parameters, including the laser power and welding speed, on hot cracking, the weld microstructure, and the concentration of alloying elements. This investigation facilitated the establishment of a hot-cracking mechanism for laser-welded AA6061 in a lap joint configuration. The primary conclusions are as follows:

The study demonstrated that specimens exposed to a high heat input (50 J/mm) exhibited substantial hot cracking, while those exposed to a moderate heat input (40 J/mm) showed minor hot cracking. In contrast, specimens exposed to a low heat input (30 J/mm) did not exhibit hot cracking, except for the specimen exposed to the highest laser power (2000 W). Generally, a decrease in the heat input corresponded to a reduction in the hot-cracking susceptibility.

Specimens with coarse grain structures showed significant hot cracking compared to those with finer grain structures. This study also noted that specimens exposed to a high heat input exhibited a coarse-grained structure, whereas those exposed to a low heat input had a finer-grained structure.

The hot-cracked regions showed significantly higher concentrations of alloying elements than the acceptable limits for AA6061 (Si: 2.3 times, Cr: 8.1 times, and Fe: 2.7 times, on average). The chances of hot cracking in welds may increase if the concentrations of these elements surpass these limits. As such, controlling the elemental composition is recommended to minimize the risk of hot cracking.

Hot-cracking initiation occurred at the weld root of the faying interface between the two specimens. We discovered that the initiation of hot cracking was influenced by various factors. These factors include the geometrical features of the faying interface acting as a pre-crack surface, strain development between dendrites resulting from a coherent network, and a high concentration of alloying elements reducing the cohesive force between dendrites. As a result, cracks are initiated at the faying interface located at the fusion boundary.

Regarding the crack growth mechanism, the crack initiated from the weld root at the faying interface and propagated along the growth direction of the columnar dendrites. This is because the intergranular region serves as a crack propagation path. As the crack reached the central region of the FZ, it propagated upward due to the cooling direction in the center area of the FZ. This crack propagation continued until the driving force for the crack diminished.

As mentioned above, samples with a high heat input or laser power condition were more vulnerable to hot cracking. Under these conditions, a higher melt pool temperature and more rapid changes in the temperature are induced. Accordingly, the likelihood of hot-cracking generation increases. In order to improve the hot-cracking issue in the laser welding of Al alloys, using dual-core fiber laser or dual-beam laser welding (using additional pre- or post-heating laser beams), which can provide gentle variation in the melt pool temperature, is considered a good approach.

Author Contributions: K.R.: formal analysis, visualization, investigation, and writing—original draft. S.K.: data curation, visualization, and methodology. J.S.: conceptualization, methodology, resources, supervision, project administration, funding acquisition, and writing—review and editing. All authors have read and agreed to the published version of the manuscript.

Funding: This research was supported by the Innopolis Foundation of Korea, funded by the Ministry of Science and ICT (grant No. 2023-SB-SB-0079), and the "Regional Innovation Strategy (RIS)" through the National Research Foundation of Korea (NRF), funded by the Ministry of Education (MOE) (2021RIS-004).

Institutional Review Board Statement: Not applicable.

Informed Consent Statement: Not applicable.

Data Availability Statement: The data will be made available upon request.

Conflicts of Interest: The authors declare that they have no competing financial interests or personal relationships that may have influenced the work reported in this study.

References

1. Davis, J.R. *Aluminum and Aluminum Alloys*; ASM International: Materials Park, OH, USA, 2001; pp. 351–461. [CrossRef]
2. ASM International. *ASM Handbook Properties and Selection: Nonferrous Alloys and Special-Purpose Materials*; ASM International: Materials Park, OH, USA, 2001.
3. Deekshant, V.; Kaushal, K. Application and use of different aluminium alloys with respect to workability, strength and welding parameter optimization. *Ain. Shams Eng. J.* **2021**, *12*, 1143–1152. [CrossRef]
4. Okon, P.; Dearden, G.; Watkins, K.; Sharp, M.; French, P. Laser welding of aluminium alloy 5083. In Proceedings of the International Congress on Applications of Lasers & Electro-Optics, Orlando, FL, USA, 14–17 October 2002.
5. Qiaoling, C.; Ruixiang, B.; Haigen, J.; Zhenkun, L.; Ning, H.; Cheng, Y. Microstructure, texture and mechanical properties of 6061 aluminum laser beam welded joints. *Mate Chara* **2018**, *137*, 269–276. [CrossRef]
6. Goncalo, P.; Sonia, M.; Andrew, D.; Stewart, W.; Supriyo, G.; Duncan, H.; Krystian, W. Laser spot welding of laser textured steel to aluminium. *J. Mate. Proc. Tech.* **2017**, *241*, 24–35. [CrossRef]
7. Enz, J.; Riekehr, S.; Ventzkea, V.; Huber, N.; Kashaeva, N. Fibre laser welding of high-alloyed Al-Zn-Mg-Cu alloys. *J. Mate. Proc. Tech.* **2016**, *237*, 155–162. [CrossRef]
8. Kuo, T.Y.; Lin, H.C. Effects of pulse level of Nd-YAG laser on tensile properties and formability of laser weldments in automotive aluminum alloys. *Mater. Sci. Eng.* **2006**, *416*, 281–289. [CrossRef]
9. Pierron, N.; Sallamand, P.; Matteï, S. Study of magnesium and aluminum alloys absorption coefficient during Nd:YAG laser interaction. *Appl. Surf. Sci.* **2007**, *253*, 3208–3214. [CrossRef]
10. Coniglio, N.; Cross, C.E. Mechanisms for Solidification Crack Initiation and Growth in Aluminum Welding. *Metal. Mater. Trans. A* **2009**, *40*, 2718–2728. [CrossRef]
11. Pi-zhi, Z.; Jing, L.; Zhi-dong, C. Effect of Si content on laser welding performance of Al-Mn-Mg alloy. *Trans. Nonfer Meta Soc. China* **2014**, *24*, 2208–2213. [CrossRef]
12. Zhou, J.; Tsai, H.L. Porosity formation and prevention in pulsed laser welding. *J. Heat. Transfer* **2007**, *129*, 1014–1024. [CrossRef]
13. Rappaz, M.; Dremaud, J.M.; Gremaud, M. A new hot tearing Criterion. *Metal. Mater. Trans. A* **1999**, *30*, 449–455. [CrossRef]
14. Kou, S. A criterion for cracking during solidification. *Act. Mater.* **2015**, *88*, 366–374. [CrossRef]
15. Liu, J.W.; Rao, Z.H.; Liao, S.M.; Wang, P.C. Modeling of transport phenomena and solidification cracking in laser spot bead-on-plate welding of AA6063-T6 alloy. Part I—the mathematical model. *Int. J. Adv. Manuf. Technol.* **2014**, *73*, 1705–1716. [CrossRef]
16. Wang, L.; Gao, M.; Zhang, C.; Zeng, X.Y. Effect of beam oscillating pattern on weld characterization of laser welding of AA6061-T6 aluminum alloy. *Mater. Des.* **2016**, *108*, 707–717. [CrossRef]
17. Niel, A.; Bordreuil, C.; Deschaux-Beaume, F.; Fras, G. Modelling hot cracking in 6061 aluminium alloy weld metal with microstructure based criterion. *Sci Technol Weld. Join.* **2014**, *18*, 154–160. [CrossRef]
18. Fabrègue, D.; Deschamps, A.; Suéry, M.; Proudhon, H. Two- and three-dimensional characterizations of hot tears in a Al-Mg-Si alloy laser weld. *Scr. Mater.* **2008**, *59*, 324–327. [CrossRef]
19. Wang, J.; Wang, H.P.; Wang, X.J.; Cui, H.C.; Lu, F.G. Statistical analysis of process parameters to eliminate hot cracking of fiber laser welded aluminum alloy. *Opt. Laser Technol.* **2015**, *66*, 15–21. [CrossRef]
20. Christian, H.; Daniel, W.; Rudolf, W.; Thomas, G. Reduction of the hot cracking susceptibility of laser beam welds in AlMgSi alloys by increasing the number of grain boundaries. *Sci. Techn Weld. Join.* **2018**, *24*, 313–319. [CrossRef]

21. Sindo, K. Solidification cracking susceptibility associated with a teardrop-shaped weld pool. *Sci. Techn Weld. Join.* **2021**, *26*, 341–347. [CrossRef]
22. Wang, L.; Wu, T.; Wang, D.; Liang, Z.; Yang, X.; Peng, Z.; Liu, Y.; Liang, Y.; Zeng, Z.; Oliveira, J.P. A novel heterogeneous multi-wire indirect arc directed energy deposition for in-situ synthesis Al-Zn-Mg-Cu alloy: Process, microstructure and mechanical properties. *Addit. Manuf.* **2023**, *72*, 103639. [CrossRef]
23. *ASTM E29-13*; Standard Practice for Using Significant Digits in Test Data to Determine Conformance with Specifications. ASTM International: West Conshohocken, PA, USA, 2013.
24. Wang, X.J.; Wang, H.P.; Lu, F.G.; Carlson, B.E.; Wu, Y.X. Analysis of solidification cracking susceptibility in side-by-side dual-beam laser welding of aluminum alloys. *Int. J. Adv. Manuf. Tech.* **2014**, *73*, 73–85. [CrossRef]
25. Dmitry, G.E.; Laurens, K. A Quest for a New Hot Tearing Criterion. *Metal. Mater. Trans. A* **2007**, *38A*, 1511–1519. [CrossRef]
26. Ravi, K.R.; Pillai, R.M.; Amaranathan, K.R.; Pai, B.C.; Chakraborty, M. Fluidity of aluminum alloys and composites: A review. *J. Alloys Comp.* **2008**, *456*, 201–210. [CrossRef]

MDPI AG
Grosspeteranlage 5
4052 Basel
Switzerland
Tel.: +41 61 683 77 34

Materials Editorial Office
E-mail: materials@mdpi.com
www.mdpi.com/journal/materials

www.ingramcontent.com/pod-product-compliance
Lightning Source LLC
Chambersburg PA
CBHW041342210926
51850CB00013B/2956
* 9 7 8 3 7 2 5 8 4 3 2 1 3 *